Settled Asbestos Dust Sampling and Analysis

Settled Asbestos Dust Sampling and Analysis

James R. Millette
Steve M. Hays

CRC Press
Taylor & Francis Group
Boca Raton London New York

CRC Press is an imprint of the
Taylor & Francis Group, an **informa** business

First published 1994 by Lewis Publishers

Published 2020 by CRC Press
Taylor & Francis Group
6000 Broken Sound Parkway NW, Suite 300
Boca Raton, FL 33487-2742

First issued in paperback 2020

ISBN 13: 978-0-367-57978-4 (pbk)
ISBN 13: 978-0-87371-948-3 (hbk)

Visit the Taylor & Francis Web site at
http://www.taylorandfrancis.com

and the CRC Press Web site at
http://www.crcpress.com

Library of Congress Card Number 94-7111

Library of Congress Cataloging-in-Publication Data

Millette, James R.
Settled asbestos dust sampling and analysis / by James R. Millette and Steve M. Hayes.
p. cm.
Includes bibliographical references and index.
ISBN 0-87371-948-4
1. Asbestos dust—Measurement. I. Hays, Steve M. II. Title.
TD887.A8M54 1994
363.17'91—dc20 94-7111
CIP

Acknowledgment

The authors gratefully acknowledge the invaluable assistance of Phyllis C. Gobbell with the writing of this book. This book would not have been possible without her skill as an editor and her determination as a coordinator.

About the Authors

James R. Millette received a B.S. in Physics from the University of Dayton, Dayton, Ohio, in 1973; a Masters of Environmental Science from Miami University, Oxford, Ohio, in 1975; and a Ph.D. from the School of Engineering, University of Cincinnati, Cincinnati, Ohio, in 1983. Dr. Millette spent 11 years with the United States Environmental Protection Agency (EPA) where he served as chairman of the electron microscope facility. While at the EPA, he designed, implemented, and directed studies concerning exposure assessment, epidemiology, and toxicology of environmental contaminants. Dr. Millette was employed for 5 years by McCrone Environmental Services where he was involved in numerous analytical projects and asbestos studies. He served as Laboratory Director and Vice President and General Manager. He is currently Executive Director of MVA, Inc., a company that specializes in the microanalysis of small particles.

Dr. Millette has authored over 50 papers and published in scientific journals. He has frequently been invited to present lectures on the analysis of asbestos fibers and serves as a co-director for the Georgia Tech course entitled: "Settled Dust Analysis: Asbestos and Other Particulates." He is a member of the American Association for the Advancement of Science, the Environmental Information Association, and the American Society for Testing and Materials (ASTM), where he serves as Vice Chair of Subcommittee D22.07 on Sampling and Analysis of Asbestos in Ambient Environments. He also serves as the coordinator for the proposed method for the analysis of asbestos in air.

Dr. Millette's proceeds from the sale of this book will be donated to the Georgia Microscopical Society, a non-profit organization which sponsors a young people's microscopy course to promote interest in the use of the microscope among school age children.

Steve M. Hays, P.E., C.I.H. graduated from Vanderbilt University, Nashville, Tennessee, in 1973 with a Bachelor of Engineering degree in Chemical Engineering. He became a registered Professional Engineer in April, 1978, and was certified as an Industrial Hygienist in June, 1990, by the American Board of Industrial Hygiene.

Mr. Hays was employed by E.I. DuPont Company from June 1973 until 1978, and was a Senior Engineer when he left DuPont to join Gobbell and Associates in September, 1978, now Gobbell Hays Partners, Inc. Gobbell Hays Partners, Inc. is an Architectural/Engineering/Environmental Consulting firm with offices in Nashville and San Francisco.

Mr. Hays served as President of NAC/The Environmental Information Association during 1992–1993. He served as President of the Consulting Engineers of Tennessee during 1991–1992 and is a member of the National Institute of Building Sciences (NIBS), Consultative Council. Mr. Hays was also a consultant to NIBS for the development of *NIBS Guidance Manual: Asbestos Operations and Maintenance Work Practices*.

Mr. Hays served as a member of the NIBS Task Force Steering Committee for the preparation of *Asbestos Abatement & Management in Buildings: Model Guide Specifications*. He has also served on the NIBS lead-based paint and radon committees. He was a member of the Environmental Protection Agency's Policy Dialogue Committee, which advised EPA on future regulation of asbestos-containing materials in public and commercial buildings and served as reviewer for EPA's *Managing Asbestos in Place* ("Green Book"), issued in July, 1990, which provides guidance for the in-place management of asbestos-containing materials. Mr. Hays also served on the EPA's 24-member negotiating committee that established the regulations for the Asbestos Hazard Emergency Response Act (AHERA), affecting all private and public school districts across the nation. Mr. Hays is a member of the seminar faculties at Georgia Tech Research Institute and The Environmental Institute, where he conducts dozens of lectures each year on asbestos, lead, and other environmental topics.

Table of Contents

CHAPTER 1

Introduction

Pieces of asbestos-containing materials (ACM's) found in buildings may be released from the in-place material by a number of mechanisms including physical disturbance, deterioration, and the natural aging of the material. Some of the released material is in the form of visible pieces of debris, some as individual fibers of asbestos which can only be seen under an electron microscope. All of the large and most of the small fiber structures and non-asbestos particles which are released eventually settle on the surfaces of ceiling tiles, light fixtures, furnishings, equipment, floors, and carpeting and become part of the normal dust. Activities which disturb the debris or dust such as maintenance or custodial tasks are likely to re-entrain the asbestos, and this re-entrainment may cause personal exposure to airborne asbestos. Analysis of settled dusts and surface particles of all sizes for asbestos is useful for qualitative and quantitative assessment and may help determine the source of the fibers. This information may be helpful in developing plans to minimize the build-up of asbestos-containing dust by controlling the source and by preventing resuspension of dust.

The study of asbestos-containing debris and settled dust is part of the general field of study of surface contamination. The term "surface contamination" is used to describe the fouling of a surface with an undesired or unexpected chemistry[1] which, for instance, in the semiconductor industry may represent any particle which prevents the fabricated part from functioning properly. In the assessment of buildings, "surface contamination" refers to the part of the dust which contains materials that are considered toxic. According to the Occupational Safety and Health Administration (OSHA) Field Operations Manual, there are a variety of reasons why surface contamination, and especially removable surface contamination, may need to be assessed.[2] One reason is that accumulated toxic materials (e.g., asbestos, lead, or beryllium) may become resuspended in air and may contribute to airborne exposures.

Although this book emphasizes asbestos in the settled dust, sections about lead and other particulate have been included because many of the technologies which are being studied in regard to various surface contaminant situations can be used in conjunction with asbestos fibers.

Although a number of published articles have dealt with the subject area of asbestos and other particulate in surface contamination[3], this book is an attempt to compile the most significant information for ready reference.

REFERENCES

1. **Fish, B.R.**, Ed., *Surface Contamination*, Pergamon Press, Oxford, England, 1967.
2. **Occupational Safety and Health Administration**, Sampling for surface contamination, *OSHA Instruction CPL 2–2.20*, Occupational Safety and Health Administration, Washington, D.C., March 30, 1984, Chapter VIII.
3. **Brackett, K.A., Seitz, S.D., and Clark, P.J.**, EPA bibliography on surface contamination, *Microscope*, 40, 164, 1992.

CHAPTER 2

Definitions and General Characteristics of Dust

DEFINITIONS

The term "settled dust" as used in conjunction with asbestos in buildings was defined in a publication in 1986 as loose fibers and particles that have collected on building surfaces.[1] However, the term "dust" has no single definition. It has a common usage among homemakers and building owners as the fine particle layer that accumulates on furnishings and must be "dusted" off. It is also used in a variety of scientific and technical disciplines to describe a collection of small particles, but seldom are these definitions the same. To further confuse the issue, the term is used to describe a collection of small particles which is both airborne and has settled or accumulated on surfaces. Although "dust" conveys the same general meaning to many people, it does not have one meaning which can be considered a single definition.

In physics, "dust" is a loose term applied to solid particles predominantly larger than colloidal size and capable of temporary gas suspension.[2] In the general geological sense, the term "dust" refers to wind transported regolith, the debris resulting from weathering.[3] In aerosol science, the term "dust" is one of several terms used to describe different types of particles comprising aerosols, microscopic sized particles dispersed in gases, often in air.[4] The term "dust" is also used to describe a dispersion of solid particles usually resulting from the fracture of larger masses of material such as in blasting, drilling, crushing or grinding operations. Whereas "mists" are a dispersion of liquid particles, many of which are visible, "fog" refers to visible aerosols of a liquid formed by condensation; "fume" is an aerosol of solid particles formed by condensation of vaporized materials; and "smoke" refers to an aerosol resulting from incomplete combustion mainly of carbon and other combustible materials. These aerosol definitions sometimes overlap because they have arisen from popular usage. Settled dust, accumulations of solid particles that have fallen out of the air, is also referred to as "accumulated dust," "surface dust," and in some cases as "surface contamination."

Although dust is considered to be composed of particles that are small, there is no universally accepted size below which particles are considered to be dust particles. A chart entitled "Characteristics of Particles and Particle Dispersoids" is frequently cited in air pollution texts[5–6] to describe the characteristics of particles. In the Technical Definitions section of this chart, the term "dust" is shown to encompass particle diameters from 1 micrometer to as large as 10,000 micrometers (1 centimeter).[7] An encyclopedia lists dust as being composed of many kinds of solid materials which have diameters less than 0.0025 millimeters (2.5 µm).[8] In geology, the term refers to dry solid matter of silt and clay size which is taken to be less than 1/16 of a millimeter (62.5 µm).[9] In a 1957 survey of airborne dust properties, the particle size range of airborne dust was considered to be from about 0.02 to 50 µm, although it was recognized that different size analysis methods resulted in different types of particle diameter values.[10] In the standard of the American Society of Heating, Refrigerating and Air-Conditioning Engineers (ASHRAE) covering ventilation for acceptable indoor air quality, "dust" is defined as an air suspension of particles (aerosol) of any solid material, usually with particle size less than 100 micrometers (µm).[11] The best definition for settled dust is found in the American Society for Testing and Materials (ASTM) test method for the collection and measurement of dust fall. Settleable particulate material is defined as any material composed of particles small enough to pass through a one (1) mm screen and large enough to settle by virtue of their weight into a container from the ambient air.[12]

Although the term "debris" is often associated with pieces of material generally larger than dust, it is not a term defined in terms of size. The United States Environmental Protection Agency (EPA) has defined the term "asbestos debris" as related to ACM in its Asbestos Hazard Emergency Response Act (AHERA) regulation as pieces of asbestos-containing building material (ACBM) that can be identified by color, texture, or composition.[13] The EPA definition also includes dust, if the dust is determined by an accredited inspector, to be ACM.

For the purposes of this book, settled dust is defined as being particulate material less than one (1) mm in its longest dimension and which is deposited on a surface. Debris is defined here as pieces of material which can be identified as to probable source by visual assessment of their color, texture, or composition.

CHARACTERISTICS OF DUST

In order to understand asbestos in surface contamination, it is useful to have some information about the general nature of dust and debris. The composition of building and household dust is a combination of particles resulting from human (and pet) activities, general degradation of building materials, and a contribution from the outdoor dust that comes in by a variety of ways including the air system, through open windows, or carried in on shoes. The relative contribution of each source will vary according to geographic location, location within the building, and activities which occur in the building, including the type of air-conditioning and ventilation system.

There are a number of sources of dust in the outdoor setting. Cosmic dust which reaches the earth's surface is mainly related to the disintegration of meteorites (black magnetic spherrules). Their contribution to the overall outdoor settled dust is small in comparison with other sources. Among other natural sources, volcanic dust can be a significant source of dust particles. Some dust particles originate in the process of conversion of gas to particulate which occurs with sulfur and nitrogen oxides in the air. Forest fires and other burning may contribute both organic and inorganic particles to the dust. Along the coastal areas, sea salts may be an important source of dust particles. Deflation (wind erosion) of soil is a main source of atmospheric dust particles larger than 2 μm. Airborne dust also contains pollens, spores and decaying vegetable material.[14]

Industrial emissions such as those from smoke stacks, stockpiles, and waste dumps may contribute to the airborne dust load. In the local area surrounding a building, the activities of demolition and construction can cause significant amounts of airborne dust. The natural degradation of exterior building materials as they are affected by the elements contribute to the components of dust. Similarly the weathering of rocks produces fine particles which in some cases may become airborne. Once put into the air, all of these particles eventually fall to the ground, but many are carried to earth in rain and snow.

Many settled dust particles are created by the deterioration of materials. Greathouse and Wessel have defined deterioration as a process of transition from a higher to a lower energy state and have also stated that physical, chemical and biological factors or agents all may play roles in promoting degradation.[15] The local environment determines which agents are present in a given spot to cause the deterioration. Physical deteriorative agents include heat, sunlight, vibration, and air which includes the abrasive agents of dust and grit in the airflow. Chemical deteriorative agents include moisture, salts, acids, alkalies, and gases. Biological agents include fungi (molds), bacteria, and insects. Moisture in the form of humidity plays an especially significant role in degradation of some materials. Small amounts of atmospheric acids are created by the combination of moisture and sulfur dioxide and moisture and carbon dioxide, which yield respectively sulfurous and carbonic acids. In general, the more water, the more and faster the degradation of most materials.[16]

Although some characterization of the particles that make up outdoor dusts and those found in industrial settings has been done, research on examining particles found in dusts from commercial buildings and residences is limited. The 1957 survey of the American Society of Heating and Air-Conditioning Engineers (ASHAE) provided some data on the characteristics of typical airborne dust particles in residences and commercial buildings in Minneapolis, Akron, Pittsburgh, and Louisville.[17] Approximately 10% (range 0–20%) of the particles were spherical, thought to represent smokes, pollen, and fly ash; 40% (range 10–90%) irregular cubical, representing mineral and cinder particles; 5% (range 0–10%) flakes, representing minerals and skin particles; 10% (range 3–35%) fibrous particles, representing lint and plant fibers; and 15% (range 0–40%) condensation flocculants, representing carbon smokes and fumes. The particle shape determinations were done by light microscopy techniques.

The first large-scale study of household dust was conducted by the Hoover Company in 1928 and involved 672 samples of carpet dirt collected from 56 cities in the United States.[18] The samples were analyzed for particle size by sieving, for percent soluble in carbon tetrachloride and percent combustible at 1800°C. An additional study was performed by the Hoover Company in 1975 with 24 samples of carpet dirt collected from homes around the North Canton, Ohio area. The results of the Hoover studies were very similar to those published in a report by the University of Nebraska in 1961.[19] The percentage of surface dirt which was retained on a #50 mesh screen (about .3 mm hole size) was approximately 20% in the 1928 Hoover report, 17% in the 1961 University of Nebraska report, and 14% in the 1975 Hoover report. The percentage of dirt which passed through a #200 mesh screen (about 0.08 mm hole size) was approximately 52% in all three reports. Approximately 50% of the dirts was combustible in all three studies. The percentage of dirt soluble in carbon tetrachloride were approximately 4% in the 1928 report, 1% in the 1961 study, and 7% in the 1975 report.

In the 1928 study, some additional work suggested that the fraction extractable by organic solvent was probably composed of mineral, animal, and vegetable oils, rubber, asphalt, tar, and resins. The fraction which was combustible at 1800°C was thought to include, in addition to the solvent extractables, moisture, hair, wool, cotton, wood, and paper. Of the remaining inorganic fraction, chemical analysis suggested minerals primarily composed of silicon, aluminum, iron, and calcium, such as quartz, feldspar, and gypsum. The data from the ASHAE and Hoover studies provide only a beginning of an understanding of what constitutes the material called "settled dust."

REFERENCES

1. **Light, E.N. and Jankovic, J.T.**, Assessment of asbestos fiber release in buildings through analysis of settled dust, *NAC Journal*, 4(4), 9, 1986.
2. **Parker, S.P.**, Ed., *Dictionary of Scientific and Technical Terms*, 4th Ed., McGraw-Hill, New York, 1989.
3. **Dietrich, R.V. and Skinner, B.J.**, *Gems, Granites & Gravels*, Cambridge Univ. Press, Cambridge, UK, 1990, 89.
4. **Reist, P.C.**, *Aerosol Science and Technology*, McGraw-Hill, Inc., New York, 1984, 2.
5. **Cooper, D. and Alley, C.E.**, *Air Pollution Control: A Design Approach*, PWS Publishers, Boston, 1986.
6. **Licht, W.**, *Air Pollution Control Engineering*, 2nd Ed., Marcel Dekker, Inc., New York, 1986, 15.
7. **Lapple, C.E.**, *SRI Journal*, 5, 94, 1961.
8. **Miller, J.E.**, Dust, *The Encyclopedia Americana*, Vol. 9, Grolier Inc.,1990, 497.
9. Dietrich, R.K. and Skinner, B.J.
10. **Whitby, K.T., Algren, A.B., Jordan, R.C. and Annis, J.L.**, The ASHAE air-borne dust survey, *Heating, Piping and Air Conditioning*, Nov., 187, 1957.
11. **American Society of Heating, Refrigerating and Air-Conditioning Engineers**, Ventilation for acceptable indoor air quality, *ASHRAE Standard 62-1989*, American Society of Heating, Refrigerating and Air-Conditioning Engineers, Inc., Atlanta, GA, 1989.
12. **American Society for the Testing and Materials**, Standard test method for the collection and measurement of dustfall (settleable particulate matter). *ASTM Standard D1739-89*, Vol. 11.03, American Society for Testing and Materials, Philadelphia, PA, 1991, 44.
13. **United States Environmental Protection Agency**, Asbestos-containing materials in schools; final rule and notice, 40 CFR, Part 763, *Federal Register*, Vol. 52, No. 210, Oct. 30, 1987, 41826.
14. **Pye, K.**, *Aeolian Dust and Dust Deposits*, Academic Press, NY, 1989.
15. Ibid.
16. **Greathouse, G.A. and Wessel, C.J.**, *Deterioration of Materials*, Reinhold Publ. Co., New York, 1954, 4,71.
17. **Whitby, K.T., Algren, A.B. and Jordan, R.C.**, ASHAE Research Report No. 1549, Size distribution and concentration of air-borne dust. *ASHAE Transactions* 61, 463, 1955.
18. **Hoover Co.**, Carpet Dirt Research, The Engineering Div., The Hoover Co., North Canton, OH, 1928, 1975 (Provided by J.N. Balough, the Hoover Co., North Canton, OH).
19. Univ. of Nebraska Report on U.S. Dirt 1961 (Provided by J.N. Balough, the Hoover Co., North Canton, OH).

CHAPTER 3

Settled Dust Methods: History and Discussion

A number of methods have been used to collect particles from surfaces for subsequent chemical analysis, and a few have been used for examination of the particles by microscopy. Generally these fall into five categories: scrape or scoop methods, adhesive tape sampling, vacuuming, wiping (wet and dry), and passive monitors (in situ dustfall collectors). Depending on the situation, one may be more useful than the others.

All five types of collection procedures have been used for asbestos, and the choice of which to use depends on the situation and the information which is desired. For most situations, the settled dust to be collected occurs in layers that are more than one particle thick. A direct examination of a thick layer of dust allows only the surface of the layer to be viewed under the microscope. In order to analyze for asbestos fibers throughout the sample, some modification of the dust must be made to provide a single layer of dispersed particulate. All of the feasible specimen collection and preparation techniques result in some modification of the dust particulate. For lightly loaded surface areas (where a layer of dust is not visible), an adhesive tape sampler may be used to collect material. For areas where the dust is visible, the scrape, vacuum, or wipe samples are most appropriate. When it is necessary to gain information about the rate of accumulation of dust, the passive monitors are most appropriate.

The information that follows is general in nature. It is intended to guide experienced field personnel and is not intended to be comprehensive in specific detail. Procedures may change over time, and the reader is encouraged to consult with the EPA, OSHA, ASTM, or other groups involved with publishing protocols to obtain the most current versions. Protection of the building environment, protection of building occupants, and the health and safety of the persons doing the sampling are not addressed. The reader is cautioned that all work should be done using good practices which will protect the building environment, the building occupants, and those doing the sampling.

SCRAPE AND SCOOP

J. Guth reported that he had begun using a sharp-edged, flat bladed screwdriver and 35 mm film canisters in his surface scrape sampling procedure in 1983 for checking surfaces which had been abated of asbestos-containing material.[1] However, the scrape and scoop method of collecting samples of settled dust for asbestos examination had already been in use for a number of years. In 1935, Hurlbut and Williams collected settled dust samples from six asbestos plants in various parts of the U.S. by scooping rafter dust into containers and characterizing the fibers by polarized light microscopy (PLM).[2] The current technique of scrape and scoop sampling involves scraping together a pile of dust by using a card (such as a credit card) and scooping the dust pile into a 35 mm film canister or similar container that can be closed tightly. If the dust layer is thick enough, it may be possible to scoop the material directly into the canister. If quantitative estimation is desired, dust must be collected from a measured area of the surface. The samples may be examined by PLM or transmission electron microscopy (TEM) for the identification of asbestos.

ADHESIVE TAPE LIFTS

A number of adhesive tape lift materials have been used to collect particles from surfaces for microscopic examination. Tape lifters for latent finger and foot prints have been common forensic tools for crime scene investigations. Gelatin lifters have also been used by the forensic community to pick up powder particles.[3] The advantage of gelatin lifters is their elasticity and low tack. Particles on uneven surfaces such as stucco can be lifted more effectively with rubber-backed gelatin lifters than with tape. The special sticky gelatin foils developed for forensic purposes have been used to detect mineral wool fibers in the indoor environment and to measure the amount of dust on a surface, using a light scattering technique.[4–5] Post-it™ notes have been used to collect particles on surfaces for scanning electron microscopic (SEM) examination to aid in determining the source of the particles.[6] Guth has reported using fingerprint tape, Scotch™ transparent tape, and dust-lift gelatinous foil films to collect samples of asbestos.[7]

Strip tape sampling methods have been employed to test abatement equipment surfaces for asbestos contamination, using commercially available cellophane tape (3M Scotch™ 810 Magic Tape™) and precleaned microscope slides.[8] The strip tape slides are prepared beforehand in a clean area by placing a clean strip of tape (2×8 cm), sticky side down, onto a clean slide. A tab is made on the end of the tape to allow for labeling and handling. The sampling process involves removing the tape from the slide, touching the piece of tape, sticky side down, onto the sampling surface several times over a 10 cm square surface area, where possible. Once the area of interest has been sampled with the tape, the tape is pressed back into the slide, sticky side down, and placed in a clean Ziploc™.

In the laboratory, each piece of tape is analyzed by bag PLM.[9] Results are expressed as a type of asbestos and percentage of asbestos relative to the total particulate on the surface sample. The percentage of asbestos in dust should not be interpreted in the way that percentage of asbestos in a building material sample is interpreted. A high percentage means that asbestos is readily found. If one fiber is found, the result is reported as *trace*. If no asbestos can be found, the result is reported as NAD (no asbestos detected).

Although adhesive samplers have been used to collect samples of particles for microscopic examination, little work has been done using adhesive samplers for quantitative asbestos analysis by TEM. One major problem with adhesive samplers for TEM study is that excessive amounts of particulate (over a single layer of dispersed dust) can overload the collection medium and prevent any success at direct analysis. If the amount of dust is visible with the naked eye on a surface, it will probably be overloaded when examined by TEM. Figures 1-5 show how a layer of dust may appear to be composed of a few complex matrices and clusters of fibers when it may instead be made of many individual fibers settling out at different times.

The other major problem with adhesive samplers for TEM analysis is the non-homogeneity of dust on a lightly loaded surface. There can be areas of the microscope grid where the material is much too heavy to be read while other areas are free of fibers.

WIPE SAMPLES

Wipe samples collected by wiping a surface with a moistened filter have been used to assess surface contamination for a number of particulate materials including lead, beryllium, and radioactive materials: thorium, plutonium, uranium, and polonium.[10–13] Wipe sampling is also known as swipe or smear sampling. OSHA has published the following instructions on how to use wipe sampling for its officers:

a. At the worksite, prepare a rough sketch of the area(s) or room(s) to be wipe sampled.
b. Put on a pair of clean impervious disposable gloves. A clean set of gloves should be used with each individual sample. This avoids contamination of the filter by the hand and the subsequent possibility for false positives, and prevents contact with the substance.

Figure 1. Hypothetical image of dust-containing asbestos fibers that settled over a period of time as it appears on a surface now. If collected and prepared by a direct procedure, the count of fibers would be 1 individual fiber and 2 complex structures (matrices). If the contribution of each layer which fell during the settling period is considered as shown in Figures 2-5, the actual fiber count is 36 fibers and 1 complex matrix.

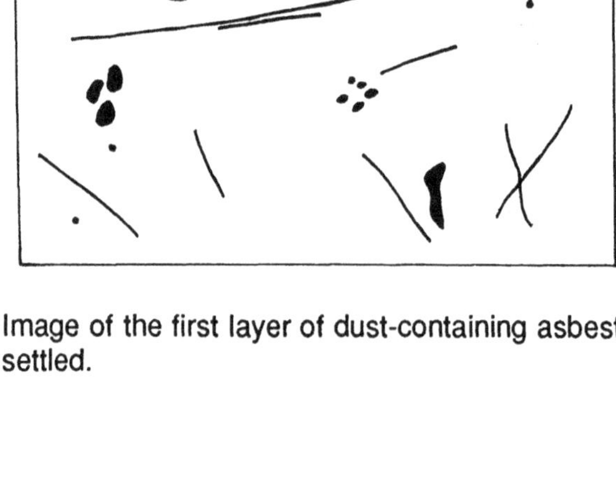

Figure 2. Image of the first layer of dust-containing asbestos fibers that settled.

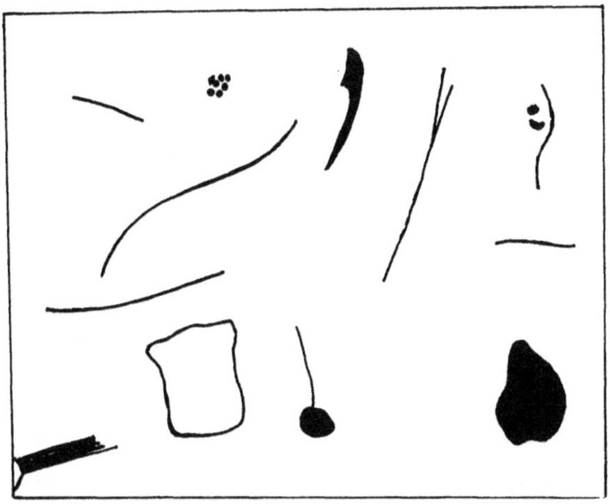

Figure 3. Image of the second layer of dust-containing asbestos fibers that settled.

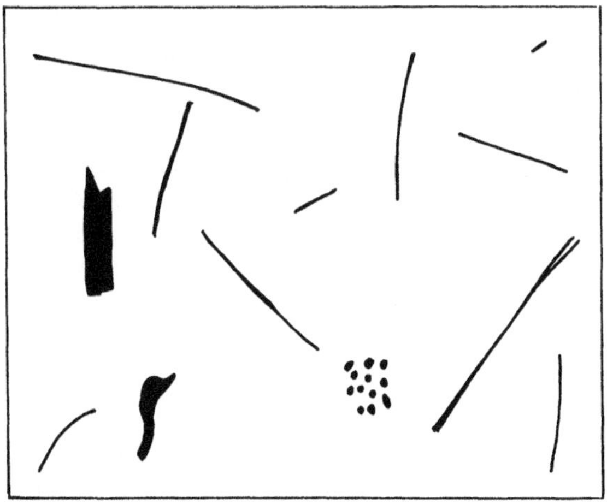

Figure 4. Image of the third layer of dust-containing asbestos fibers that settled.

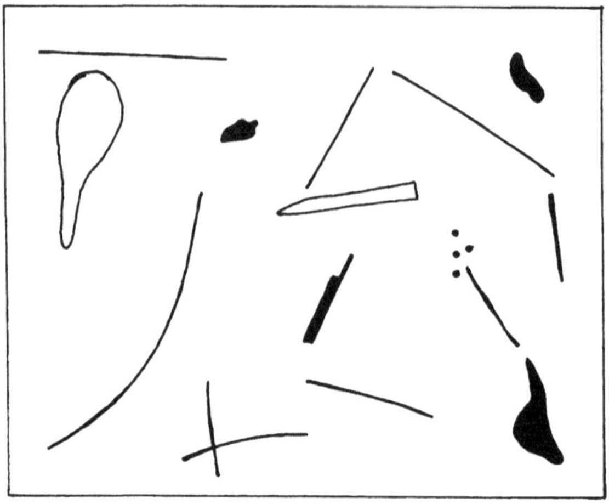

Figure 5. Image of the fourth layer of dust-containing asbestos fibers that settled.

c. Withdraw the filter from the vial. If a damp wipe sample is desired, moisten the filter with distilled water (or other solvent as recommended in Appendix A, the Chemical Information Table).
d. Wipe approximately 100 cm^2 of the surface to be sampled.
e. Without allowing the filter to contact any other surface, fold the filter with the exposed side in, then fold it over again. Place the filter in a sample vial, cap the vial, number it, and place a corresponding number at the sample location on the sketch. Include notes with the sketch giving any further description of the sample (e.g., "Fred Employee's respirator, inside," "lunch table," etc.).
f. At least one blank filter treated in the same fashion, but without wiping, should be submitted for each sampled area.[14]

In one study, commercially available wipe sampling filters called Smear Tabs were evaluated for assessing elemental surface contamination in various locations in home, office, research laboratory, and outdoor environments.[15] The study concluded that water-wetted filter samples collect about one-half the removable contamination on a wide variety of surfaces and that the elemental concentrations of surface dust can be determined to within slightly better than an order of magnitude.

Wipe Samples for Polychlorinated Biphenyls (PCB's)

According to the U.S. EPA, when a spill of polychlorinated biphenyls is cleaned up, the area must be sampled by wipe testing, and the cleaned area must meet a numerical surface standard of less than 10 micrograms of PCB's per 100 square centimeters for all indoor residential surfaces other than vault areas.[16] For PCB's on solid surfaces, a standard-size template (10 cm × 10 cm) is used to delineate the area of cleanup; the wiping medium is a gauze pad or glass wool of known size which has been saturated with hexane. The wipe test must be performed quickly after the hexane is exposed to the air because the hexane evaporates rapidly.

Wipe Samples for Lead

For lead, the Department of Housing and Urban Development (HUD) has described a specific method for wipe sampling. In this procedure, the collector uses wipes consisting of a commercial variety (baby wipes) moistened with a non-alcohol wetting agent.[17] A one square foot plastic template should be used when samples are taken from the floor. When window sills and wells are sampled, the length and width of the area are measured in inches. To collect the sample, place a wipe flat on the surface being sampled. Rub the wipe flat on the surface to be sampled. Rub the wipe in an "S" pattern once over the entire measured area, taking care not to scrub. Fold the wipe in half, folding the dust into the wipe, and rub once over the surface again at a 90 degree angle to the first "S." Fold the wipe and place it in a tube (such as a 50 ml polypropylene centrifuge tube; for example, Fisher Scientific No. 05-500-20C). Mark the tube with the sample number, location, and surface type (sill, well, floor).

Wipe Samples for Asbestos

For asbestos, a wet wipe procedure similar to the lead procedure was developed for areas where microvac sampling could not be used. Unfortunately, the commercial variety of baby wipes were found to contain too many particles to be useful as a cloth medium for asbestos. A study of 7 commercially available cloths was done by MVA, Inc., to see which had the least amount of particles. Of the cloths, the Tex-Wipe™ Alpha 10 was found to be superior to Kimberly-Clark Crew 2, Chiffons, J-Cloth, Chubs, Spiffits, and Tucks.

The standard procedure for wet wipe dust sampling for asbestos includes moistening a Tex-Wipe™ cloth with distilled water, marking off an area of 100 square centimeters or other measured size, and, with an "S" type wiping motion, sampling the area twice. The cloth is then placed in a plastic Ziploc™ bag which has been marked with a sample number, location, and date.

MICROVAC SAMPLES

The use of an air sampling cassette as a vacuuming collection device for surface contamination of asbestos was reported by Carter in 1970.[18] He collected asbestos fibers from work clothing and analyzed the filters by PCM. He reported the levels of asbestos as fibers per cm^2 of the original cloth material.

An illustration of a microvacuuming tool developed in 1972 by Walter McCrone to collect particles off surfaces for study under the microscope can be found in *The Particle Atlas*.[19] It is not known whether this tool has been used to collect settled dust samples from buildings for asbestos.[20]

Light and Jankovic field tested the use of an air sampling cassette to collect settled dust in buildings for asbestos analysis by PLM, beginning in 1984 for the West Virginia Department of Health.[21] Their collection device included a personal sampling pump set on maximum flow rate (2–4 lpm), a mixed cellulose ester (MCE) filter to trap particles, and a plastic air sampling cassette (37mm or 25mm) to contain the dust. They also used a nozzle on the end of the cassette to maximize suction. For this investigation, vacuuming of dust at each collection site continued until an easily visible accumulation of material had collected in the cassette. The dust was scraped out with a scalpel and subjected to a standard PLM analysis for asbestos. According to their report, 89 settled dust samples were collected in 28 buildings containing asbestos materials in a variety of conditions. Thirty-five percent of the samples were positive and appeared to correlate with observed or historical building conditions. The majority of sample areas where asbestos fibers were found in dust were located directly beneath ACM's. Positive sites also included areas where uncontrolled removal or renovation work had occurred.

The quantitative microvacuuming technique which uses a modified air sampling cassette to vacuum a sample of dust from a surface to be analyzed by TEM may have been first used in the 1980's by R. Hatfield in a project which required a non-destructive technique to collect dust for asbestos analysis from the furnishings in a church. Guth reported in 1988 having used a microvacuuming method in 1981, but it was not reported that he used

TEM to analyze the samples.[22] In 1989, a meeting was held at the U.S. EPA Research Center in Cincinnati in which a number of asbestos analysts from around the country discussed microvac dust sampling, specimen preparation, and analysis by TEM. Out of the meeting, based on the microvac methods used by various laboratories, came a draft EPA test method for the microvac technique. A version of this method was eventually submitted to the ASTM Committee D-22 on Sampling and Analysis of Atmospheres for consideration as a standard method. A description of the TEM method used by McCrone Environmental Services, Inc. was published in 1990.[23] A similar vacuum apparatus has been developed by researchers at the University of Cincinnati for sampling settled dust for lead. This sampler and procedure is referred to as the dust vacuum method (DVM) and involves the use of a personal air sampling pump operating at a nominal flowrate of 2.5 liters/min., fitted with Tygon tubing. The tubing was connected to a 37-mm cassette housing a mixed cellulose ester (MCE) filter with a pore size of 0.8 µm. The inlet of the cassette was fitted with a specially made nozzle which consists of a round plastic tube crimped into a small slot at one end. Research indicates that the slotted nozzle can become blocked with dust.[24]

Samples which are collected by microvacuuming are commonly referred to as microvac samples or microvac dust samples. The authors' preferred collection method is by vacuuming a 100 square centimeter area (or other known area) with a membrane filter air sampling cassette and a vacuum pump. The sampling assembly consists of a 25-mm or 37-mm diameter MCE filter contained in a three-piece standard AHERA air cassette with a 2.54-cm (one inch) piece of tubing (6.35-mm internal diameter) attached to the face cap as a nozzle. The end of the nozzle is cut at 45 degrees. The cassette is connected to a personal sampling pump with flexible tubing. The pump with cassette assembly attached is calibrated to 2 L/min. The flow rate of the pump and the internal diameter of the nozzle may vary as long as the air velocity at the inlet is 100 (± 10) cm/s. The 100-cm^2 area is vacuumed by moving the filter cassette nozzle across the area for either a known period of time (usually 2 minutes) or until there is no visible dust remaining in the area. At the end of the sampling, the cassette is turned upward before shutting off the power to the pump. The nozzle is plugged and the cassette end sealed with an end plug. The assembly is appropriately packaged to prevent separation of the nozzle from the cassette and sealed in a Ziploc™ plastic bag.

In order to be able to prepare a sample of dust which might be many layers thick, the microvac method uses the particle dispersion techniques developed over the years for the analysis of asbestos in drinking water. This procedure allows for the dilution of high concentrations of asbestos which would not be countable with any direct procedure.

Some studies have determined precision and accuracy data for the analysis of asbestos in dust collected by the microvacuuming technique. Six microvac samples using open faced cassettes were collected from 10-cm by 10-cm sections of an artificially contaminated carpet for the EPA-sponsored study of cleaning methods for the removal of asbestos fibers from carpet.[25] The calculated coefficient of variation (CV) for the samples was 166%. In another study, the CV for four samples of dust collected from random areas on an office building carpet using the open cassette microvacuuming technique was 44%, and the CV for 12 samples of dust collected from non-carpet areas of a state office building using the open cassette microvacuuming technique was 66%.[26] See data in Table 1. In a third study, side-by-side samples were collected from the tops of light fixtures in two different buildings using the microvacuuming technique with a nozzle plastic tube extension. The data are shown in Table 2. The CV's were 19% and 40%. The samples were prepared using a sonication time of 10 minutes.

Some additional procedures have been proposed as modifications of the microvac procedure. One procedure which involved using a set of sieves of different mesh sizes was proposed to allow different size fractions of the dust to be examined. A test of the additional sampling procedure was conducted with a sample of National Institute of Environmental Health Sciences (NIEHS) chrysotile, referred to in the study as "Short-Range". The NIEHS Short-Range chrysotile was originally purchased from the Union Carbide Corporation, Niagara Falls, New York, for an animal feeding study. Union Carbide referred to the material as COF-25. This material contains fibers ranging from 0.088 to 51.1 mm in length and 0.019 to 1.57 mm in width. The median length is 0.66 mm, and 90% of the fibers are under 7.8 mm in length.[27] This is a sample, therefore, of single short free asbestos fibers, most of which should pass through all sieves larger than 45 µm.

A sample of the Short-Range chrysotile was distributed as a dust and pulled through a stack of sieves (1000 µm, 500 µm, 106 µm, and 45 µm) placed before the filter in a microvac sampler. The microvac sampler was used following the protocol with a pump running at 2 l/min. Most of the chrysotile dust was caught between the 106 µm and 45 µm sieves. Very little made it to the final filter. Some forensic science texts confirmed that this result is in line with their experiences with forensic examination of soils. Saferstein states: "It is almost always desirable to determine size distribution of soil by sieving in a liquid, usually water. Dry sieving of the entire sample is generally unsatisfactory because the small particles tend to cluster."[28]

Problems were also noted with wet sieving of fibrous dusts after being collected by the microvac sampler. To allow for the analysis of the different

Table 1. Microvac dust data from similar locations.[26]

PROJECT	SAMPLE	CARPET DUST FIBERS/sq.ft.	DUST FIBERS/sq.ft.
Office	1889-31	1.70E+04	
	1889-32	4.10E+04	
	1889-33	5.90E+04	
	1889-34	4.10E+04	
State Office	2771-1		2.60E+07
	2771-2		13.0E+07
	2771-3		15.0E+07
	2771-4		15.0E+07
	2771-5		8.00E+07
	2771-6		.45E+07
	2771-7		7.60E+07
	2771-8		13.0E+07
	2771-9		4.70E+07
	2771-10		9.90E+07
	2771-11		1.40E+07
	2771-12		4.70E+07

size fractions of dust required that the sieving process retain only those particles of a specific size range on the various mesh sieves. Komar & Cui, as well as Wang & Komar, have pointed out that sieving involves not only grade selection but also grain-shape selection.[29–30] The shape of the particle is important, as well as the size of the particle. This is especially true for fibers. If a sample of dust contained only free asbestos fibers less than 45 µm in length, it is expected that essentially no particles would be found in the different wet sieved fractions and that all fibers would make it through to the final filter.

Table 2. Precision data dust analyses, 1990. (Samples collected side by side from light fixtures under fireproofing.)

PROJECT	SAMPLE	DUST STR./sq.cm.
Office	T01	94,000
	T02	130,000
	T03	85,000
	T04	110,000
Office	T10	320,000
	T11	480,000
	T12	185,000

A sample of the Short-Range chrysotile weighing 0.03187 grams was suspended in a solution of 50/50 alcohol (ethyl) and water. The suspension was mixed in a sonic bath for 3 minutes and then poured through a stack of sieves (1000 µm, 500 µm, 106 µm, and 45 µm) placed above a filter funnel which contained a preweighed 0.2 µm pore size polycarbonate filter. The liquid would not flow through all the sieves by gravity. Therefore, suction was applied. A spray bottle of 50/50 alcohol/water was used to try to wash the material through the screens. After filtration, the filter was dried and weighed. The individual screens were examined visually and under a stereo microscope.

The weight of the polycarbonate filter was unchanged. No significant amount of chrysotile made it through the sieve screens to the polycarbonate filter. Most of the Short-Range chrysotile fibers were caught on the last two screens (106 µm and 45 µm). However, some of the Short-Range chrysotile was retained on the 500 µm sieve.

It is concluded from this test that the particle retention of the various screen meshes does not represent fibers greater in size (length or diameter) than that particular mesh size. Additional research must be done before using sieves to fractionate dust samples containing fibrous materials.

Another approach to collecting sized fractions of settled dust for asbestos is use of a vacuuming procedure with a cyclone system attached. This type of vacuum sampler has already been developed for collecting settled dust for lead. The High Volume Small Surface Sampler (HVS-3) developed at the Kennedy-Krieger Institute in Baltimore uses a cyclone and a high flowrate (estimated at 16 liters per minute). Dust samples are collected in tared teflon vessels, which are subsequently weighed to determine the weight of the dust in an area of known size. The HVS-3 system has been evaluated in controlled testing. The investigators found that there were several difficulties with the cyclone approach. One concern was that the HVS-3 uses a Dirt Devil mini-vacuum cleaner that requires AC current. This reduces the portability of the device. The flowrate for the vacuum cleaner is not specified and is not determined for each application. It is likely that the flowrate changes somewhat on surfaces presenting differing pressure drops, such as carpeting versus vinyl floors. The investigators reported that one major difficulty with the cyclone type sampler is that the cyclone must be broken down and cleaned after each sample is collected, a process that requires 5–10 minutes. Some areas of the cyclone system are difficult to reach and can only be cleaned by a thorough washing that is not feasible in the field. There was also evidence that the cyclone does not capture small particles and, therefore, would miss the fibers of interest.[31]

BULK CARPET SAMPLES

The determination of the amount of asbestos in the pile or on the surface of carpeting is a special case of collecting samples of settled dust. Samples of the surface dust can be collected by the microvac technique. However, to assess the amount of asbestos in the whole carpet, in the pile as well as on the surface, a technique in which a portion of the carpet is shaken in water to release all the fibers from the pile must be used.[32] Samples of carpet are collected by cutting a piece (usually 10 cm by 10 cm) from the carpet with a razor blade or utility knife and placing it in a wide-mouth polyethylene jar or Ziploc™ bag. (See "Methods for the Analysis of Carpet Samples for Asbestos" in Appendices for a complete discussion of this method.)

PASSIVE MONITORS (IN SITU DUSTFALL COLLECTION TINS)

Passive dust sampling using "dust fall jars" or other containers to collect dust over a known period of time has been standard procedure in environmental dust studies for many years. In the 1969 text by Rossano, the section on air sampling and analysis equipment states: "The rate of deposition of particulate matter from the air is measured by means of dustfall jars."[33] Similar descriptions are also given in other texts.[34–35] Generally, total particulate material is determined and the results expressed in terms of amount per area per time (often as tons per square mile per month). In some cases, extended chemical analyses of the collected material are undertaken. Rossano also lists other methods for evaluating the character and extent of dustfall to include the use of trays and glass slides coated with an adhesive or sticky paper tape.

During the 1957 ASHAE airborne dust survey, dust fall was measured using standard 25 × 75 mm microscope slides resting on small wood blocks.[36] Before use, these slides were carefully cleaned with detergent and water. The ASHAE survey used a photomultiplier measurement of light scattered by the dust deposits on the microscopic slides. Based on the principle that an optical density value could be calculated from the amount of light transmitted through a dirty slide verses the amount transmitted through a clean slide, the technique had a sensitivity sufficient to measure the dust fall accumulated in one day in a clean location with an accuracy of about 20%. The ASHAE survey included 55 measurements of dust fall in Minneapolis, mostly from inside offices. The dust fall accumulation rate for samples collected outside buildings was found to be five times the rate for inside samples. The average dust fall rate found for inside samples was 1.4 tons per square mile per month, or, in terms more reasonable for indoor considerations, 1.6 $\mu g/cm^2/day$.

Yamate described a method for the determination of the asbestos fallout rate that was developed in conjunction with the U.S. Consumer Product Safety Commission.[37] This procedure involved the use of 37-mm or 47-mm polycarbonate filters centered onto the bottom of a plastic petri dish as the collection medium. The petri dish and filter were lightly coated with carbon to minimize static charges and then transported covered to the collection site. At the site, the cover was removed and the filter exposed for a defined period of time, usually a minimum of three weeks. After the defined period of time had transpired, the cover was replaced and the petri dish was very carefully hand-carried to a nearby location for carbon coating of the filter. The filters were analyzed by electron microscopy, both scanning and transmission, using direct preparation techniques for airborne samples.

Segrave reported using a passive monitoring procedure (in situ dust sampling) in which the collection tins are washed out and treated as water samples, thus eliminating the need for hand-carrying the filters to a local carbon-coating site and risking particle loss from the surface.[38]

The authors use passive monitoring tins to gain information about the rate of accumulation of asbestos fibers on a surface inside a building. Similar information can be obtained by cleaning an area of asbestos fibers and then collecting a microvac dust sample from the same area at a later time. Forensic evidence tins are used by the authors for passive monitoring because they provide a convenient way to establish a clean surface at a precise starting time. Passive monitors require no auxiliary power and can collect particles settling out of the air over a long time period and thereby capture fallout from episodic events of short duration and high particulate concentration. In addition, airborne particles which occur in the air in concentrations too dilute to measure with air sampling procedures that use filters and air pumps may accumulate in the passive monitors to the point where they can be measured.

The authors' current procedure involves placing the forensic tins above ceiling tiles or in other areas where they are unlikely to be disturbed and leaving them from periods ranging from 1 month to 2 years. The tins are retrieved, sealed, and sent to the laboratory where they are washed out with particle free water, filtered, and analyzed by TEM. Results are reported in terms of asbestos structures per cm^2 per day ($s/cm^2/day$).

The advantages of passive monitoring over microvac sampling are: 1) the time interval of dust accumulation is known; 2) the method provides information about particles which settle out of the air over a long period of time; and 3) power is not required at the sampling site.

REPORTING OF RESULTS

The usual way of reporting results from the microvac, wipe sampling, or passive monitor test is in terms of asbestos structures per unit area. This is the same way that the results of analyses of other surface contaminants are reported. PCB's are reported in terms of $\mu g/100\ cm^2$. Uranium, transuranics, and beta-gamma emitters are reported in terms of $dpm/100\ cm^2$. The results of lead samples when collected by wipe sampling are given in terms of $\mu g/ft^2$. In each case, the surface contaminant is reported in terms of an amount per unit area. Asbestos is generally reported in terms of s/cm^2 or s/ft^2.

There have been certain studies in which the results of the analyses of surface dust for asbestos have been reported in terms of a weight percent, as is used for bulk samples of asbestos-containing materials. However, surface dust and bulk materials are very different. The EPA defined asbestos-containing materials as products containing 1% asbestos as a way to distinguish building materials to which asbestos had been intentionally added. The percentage of asbestos in ACM does not change over time, but the percentage of asbestos in the dust may change. If a surface dust contains one million asbestos structures per square centimeter and the asbestos in the dust constitutes 5% of the dust particles, adding non-asbestos particles to the dust may change the percentage of asbestos to less than 1% but does not change the fact that there are still one million structures per square centimeter on the surface.

The inherent problem with reporting asbestos surface values on a percentage basis is evident in the following example in which two surface areas were sampled before a cleaning procedure began. A very clean surface which contained only a few asbestos particles was reported as 100% asbestos, while a surface which had many asbestos fibers but also various other material particles was reported as less than 1%.[39] The asbestos Operations and Maintenance (O&M) program designer who had to decide how to deal with the cleaning of the two surfaces was at a loss. The surface which contained the most asbestos fibers was not evident in the reports. The concentration of the asbestos in the dust is not useful when trying to establish how much asbestos is present.

REFERENCES

1. **Guth, J.,** A guide to surface sampling methodology, unpublished notes presented at the Nat. Asbestos Council., Sept. 19, 1988, Boston.
2. **Hurlbut, C.S. and Williams, C.R.,** The mineralogy of asbestos dust, *J. Indust. Hyg.*, 17, 289, 1935.
3. **Carrick, B.P. and Carick, M.F.,** *Lightning Powder Company Catalog, Lightning Power Co.*, Salem, OR, 1, 9, 1992.
4. **Schneider, T., Peterson, O.H., Eriksen, P., Vinzents, P., and Hansen, B.,** A simple method for the measurement of dust on surfaces and the effectiveness of cleaning., *Environ. Internat.*, 15, 563, 1989.
5. **Schneider, T.,** Man-made mineral fibers and other fibers in the air and in settled dust, *Environ. Int.*, 12, 61, 1986.
6. **Millette, J.R., Brown, R.S., Barnett, J., and Mount, M.D.,** Scanning electron microscopy of Post-it™ notes used for environmental sampling, *NAC J.*, Spring, 32, 1991.
7. **Guth, J.**
8. **Beaulieu, H.J.,** Asbestos contamination of abatement equipment surfaces, *Appl. Occup. Environ. Hyg.*, 6(7), 583, 1991.
9. **Nichols, G.,** Scotch Magic tape - an aid to the microscopist for dust examination, *Microscope*, 33, 247, 1985.
10. **Chavalitnitikul, C. and Levin, L.,** Laboratory evaluation of wipe testing based on lead oxide surface contamination, *Am. Ind. Hyg. Assoc. J.*, 45(5), 311, 1984.

11. **Mitchell, R.N. and Eutsler, B.C.**, A study of beryllium surface contamination and resuspension, in *Surface Contamination*, B.R. Fish, Ed., Pergamon Press, Oxford, England, 1967, 349.
12. **Royster, G.W. and Fish, B.R.**, Techniques for assessing "removable" surface contamination, *Surface Contamination*, B.R. Fish, Ed., Pergamon Press, Oxford, England, 1967, 299.
13. **Saxby, W. N. and Hole, J.A.**, Practical aspects of surface contamination control at A.W.R.E., in *Surface Contamination*, B.R. Fish, Ed., Pergamon Press, Oxford, England, 1967, 209.
14. **Occupational Safety and Health Administration,** Sampling for surface contamination, *OSHA Instruction CPL 2–2.20*, Occupational Safety and Health Administration, Washington, D.C., March 30, 1984, Chapter VIII.
15. **Litchtenwalner, C.P.**, Evaluation of wipe sampling procedures and elemental surface contamination, *Am. Ind. Hyg. Assoc. J.*, 53(10), 657, 1992.
16. **U.S. Environmental Protection Agency**, 40 CFR, Part 761, Sub Part G, PCB spill clean-up policy, 1991.
17. **U.S. Department of Housing and Urban Development**, Lead-based paint: interim guidelines for hazard identification and abatement in public and Indian housing, Washington D.C., September 1990.
18. **Carter, R.F.**, The measurement of asbestos dust levels in a workshop environment, United Kingdom Atomic Energy Authority A.W.R.E. Report No. 028–70, Aldermaston, UK, 1970.
19. **McCrone, W.C. and Delly, J.G.**, *The Particle Atlas*, 2nd Ed, Ann Arbor Science Publishers, Inc., Ann Arbor, 1973, 231.
20. **McCrone, W.C.**, personal communication, 1992.
21. **Light, E.N. and Jankovic, J.T.**, Assessment of asbestos fiber release in buildings through analysis of settled dust, *NAC Journal*, 4(4), 9, 1986.
22. **Guth, J.**
23. **Millette, J.R., Kremer, T., and Wheeles, R.K.**, Settled dust analysis used in assessment of buildings containing asbestos, *Microscope*, 38, 215, 1990.
24. **Jacobs, D.E., Menrath, W.B., Succop, P.A. et al,** A comparison of five sampling methods for settled lead dust: a pilot study, Draft Report, The National Center for Lead-Safe Housing, June 15, 1993.
25. **Kominsky, J.R., Freyberg, R.W., Chesson, J., Cain, W.C., Powers, T.J., and Wilmoth, R.C.**, Evaluation of two cleaning methods for the removal of asbestos fibers from carpet, *Am. Ind. Hyg. Assoc. J.*, 51, 9, 500, 1990.
26. **Millette, J.R.**, Some comments on settled dust sample analysis by transmission electron microscopy, Nat. Asbestos Council Technical Conf., Boston, Sept. 22, 1988.

27. **McConnell, E.E., Shefner, A.M., Rust, J.H., and Moore, J.A.**, Chronic effects of dietary exposure to amosite and chrysotile asbestos in Syrian Golden hamsters, *Environ. Health Perspect.*, 53:11, 1983.
28. **Saferstein, R.**, *Forensic Science Handbook*, Prentice-Hall, Inc., Englewood Cliffs, New Jersey, 1982.
29. **Komar, P.D. & Cui, B.**, The analysis of grain-size measurements by sieving and settling-tube techniques, *J. Sedim. Petrol.*, 54:603, 1984.
30. **Wang, C. & Komar, P.D.**, The sieving of heavy mineral sands, *J. Sedim. Petrol.*, 55:479, 1985.
31. **Jacobs, D.E., Menrath, W.B., Succop, P.A., Cohen, R., Clark, C.S., Bornschein, R.L., et al,** A comparison of five sampling methods for settled lead dust: a pilot study, Draft Report, The National Center for Lead-Safe Housing, June 15, 1993.
32. **Millette, J.R., Clark, P.J., Brackett, K.A., and Wheeles, R.K.**, Methods for the analysis of carpet samples for asbestos, *Environ. Choices Tech. Supplement*, 1(2), 21, 1993.
33. **Rossano, A.T.**, *Air Pollution Control: Guidebook for Management*, Environmental Science Service Div., E.R.A. Inc., Stamford, CT, 1969.
34. **Sproull, W.T.**, *Air Pollution and Its Control*, Exposition Press, New York, 1970.
35. **McMahon, H.E.**, *Test Procedures and Methods in Air Pollution Control*, Engineering and Research Div., Air Poll. Control Dist. of Los Angeles, 1971.
36. **Whitby, K.T., Algren, A.B., Jordan, R.C., and Annis, J.L.**, The ASHAE airborne dust survey, *Heating, Piping and Air Conditioning*, Nov., 1957, 187.
37. **Yamate, G.**, Method to determine fiber release rates from asbestos containing materials in an indoor environment, *ECON: Environmental Contractor*, 14, June 1990.
38. **Segrave, A.M.**, In-situ dust sampling protocol, *Asbestos Issues*, Aug., 1990, 42.
39. **Millette, J.R.**, unpublished notes.

CHAPTER 4

Field Guide for Asbestos Dust Sampling

This chapter is designed to serve as a quick reference for the person in the field who is asked to collect samples of surface dust for subsequent analysis for asbestos. The techniques described in this section can also be used to collect samples of settled dust for other materials of interest. Depending on the specific information desired and the specific situation in the building, it may be necessary to modify the sampling procedures. Documentation of how the sample was collected is generally important in interpreting the results of analysis.

The sampling procedures presented herein have been used to sample surface areas such as the tops of ceiling tiles, heating and air conditioning duct work, the tops of bookcases and other office furniture, exit signs, window ledges, tapestries, picture frames, and carpets. Most horizontal surfaces can be sampled with a least one of these techniques. Rough surfaces such as broomfinished concrete will cause problems in sampling for the adhesive methods, the scrape and scoop, and the wet wipe methods. However, if the dust layer is thick enough, some material will be collected from the surface of the layer.

The information that follows is general in nature. It is intended to guide experienced field personnel and is not intended to be comprehensive in specific detail. Procedures may change over time, and the reader is encouraged to consult with the EPA, OSHA, ASTM, or other groups involved with publishing protocols to obtain the most current versions. Protection of the building environment, protection of building occupants, and the health and safety of the persons doing the sampling are not addressed. The reader is cautioned that all work should be done using good practices which will protect the building environment, the building occupants, and those doing the sampling.

SCRAPE AND SCOOP

Intended Use

A quick collection procedure for determining the presence and type of asbestos. In some cases, it can be used to determine a quantitative value for the amount of asbestos present.

Figure 1. Film canister and credit card used in scrape and scoop sampling of settled dust.

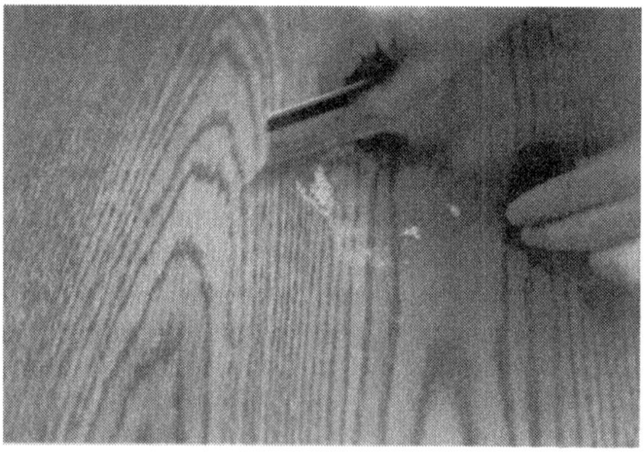

Figure 2. Scrape and scoop sampling step A: Scraping dust and debris with a credit card.

Materials

A card, such as a credit card
A 35 mm film canister or other tight-fitting container
Sample labels

Figure 3. Scrape and scoop sampling in step B: Scooping dust and debris into the canister.

Collection Procedure

1. Scrape together a pile of dust using a card (such as a credit card) and scoop the dust pile into a 35 mm film canister. (If possible, scrape the dust from a known area or surface.)
2. Mark a sample number, location, and date on the canister.
3. Ship to the laboratory for analysis. Specify polarized light microscopy analysis of the material. In some cases it may be prudent to specify transmission electron microscopy analysis. Check with the laboratory to determine what information can be determined by the two methods.

ADHESIVE TAPE LIFTS

Intended Use

A quick collection procedure for determining the presence and type of asbestos. In some cases, it can be used to determine a quantitative value for the amount of asbestos present.

Materials

An adhesive tape or roll of cellophane tape
Ziploc™ bags or other tight-fitting containers
Sample labels

Collection Procedure

1. Choose a location which has a light loading of settled dust or particulate.
2. Press adhesive sampler such as cellophane tape on material to be sampled one time. Multiple processes are not recommended under this procedure. Caution: Some surfaces may be harmed by cellophane or other adhesive samplers. Test the interaction of the adhesive and the surface material before taking samples.
3. Secure adhesive sampler by placing cellophane tape on glass microscope slide.
4. Mark the sample number, location, and date on a plastic Ziploc™ bag and place the adhesive tape lift sample in the bag. Place an unused adhesive tape lift sampler in a bag and mark as a field blank. Ship to the laboratory for analysis. Specify PLM analysis.

POST-IT™ NOTE SAMPLERS WITH CARBON TAPE

Intended Use

A quick collection procedure for determining the presence and type of asbestos and characterizing particles in the dust. It is not used to determine a quantitative value for the amount of asbestos present.

Materials

Standard Post-it™ notes to which a piece of carbon tape or carbon sheet has been added

To make a Post-it™ note sampler with a carbon tape collection portion, turn the Post-it™ note upside down so that the adhesive section is at the top. Place a one-half inch wide piece of carbon tape (available from microscopy supply distributors) below the Post-it™ note adhesive section. The carbon tape is double-sided tape which comes with plastic sheets on both sides.

Figure 4. Post-it™ note sampler with carbon tape piece. A small plastic bag is used to hold the sampler on its way to the field. Several samplers may be kept in a larger bag until needed. Step A: remove the plastic strip from the carbon tape.

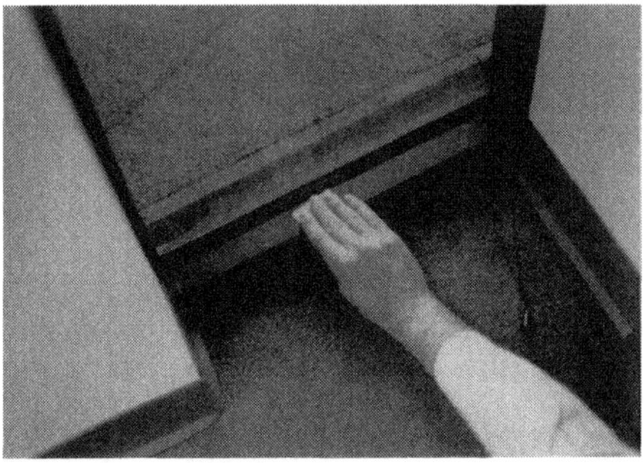

Figure 5. Post-it™ note sampler Step B: Press sampler on material of interest.

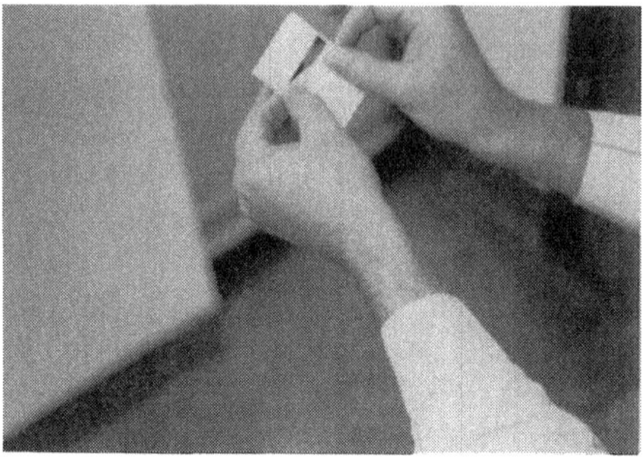

Figure 6. Post-it™ note sampler Step C: Covering carbon tape with plastic strip.

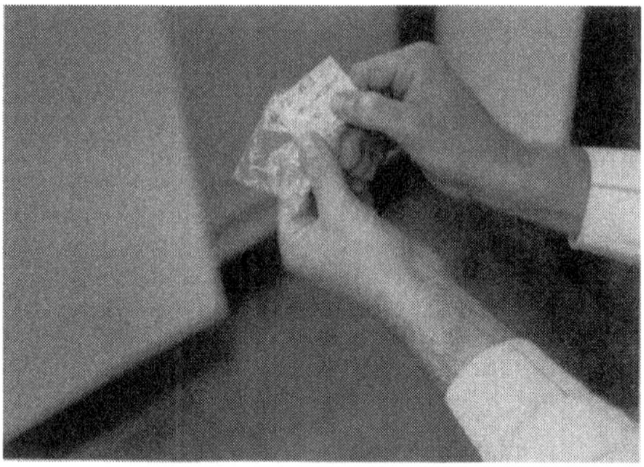

Figure 7. Post-it™ note sampler Step D: Fold the sampler over and place in a small plastic bag.

Peel one plastic sheet off, and position the carbon tape so that one side is attached to the Post-it™ note paper and the other side is covered with its plastic sheet. Fold the Post-it™ note in the middle and attach the adhesive section to the bottom of the note. Place the sampler in a plastic bag to be taken into the field. In the field, unfold the sampler and remove the plastic covering the carbon tape. Press the sampler on the area of interest and collect particles. Replace the plastic sheet on the carbon tape and refold the Post-it™ note. Place it in a plastic bag.

Ziploc™ bag or other tight-fitting container, in which folded notes are placed

Sample labels

Collection Procedure

1. Remove Post-it™ note from Ziploc™ bag and unfold, exposing sticky side of note and carbon tape.
2. Remove white protective strip from carbon tape.
3. Place sticky side of carbon tape on surface to be sampled, using three fingers. If the area is clean, note may be lifted and replaced several times. If the area is extremely dusty, a scoop sample may be collected to remove excess dust. Caution: The carbon tape or sheet sampling area may damage some surfaces. A test should be made of the surface to be sampled beforehand.
4. Replace the protective white strip over the carbon tape. If it does not stick, place it on top of the particulate.
5. Fold Post-it™ note back to its original position.
6. Write the sample number and other information directly on the note and record on the sample log sheet.
7. Repackage in Ziploc™ bag and ship to laboratory for analysis. Specify scanning electron microscope analysis of particles.

MICROVAC SAMPLING

Intended Use

A collection procedure for determining the presence and type of asbestos and a quantitative value for the amount of asbestos present.

Materials

A microvac sampler (See Chapter 3, page 16)
Sample Labels
Ziploc™ bag

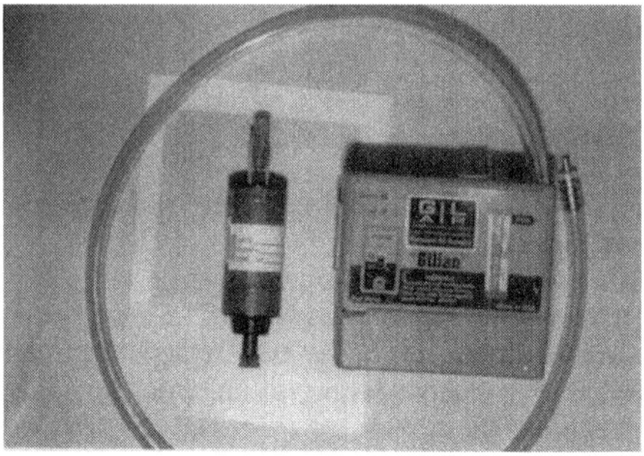

Figure 8. Personal air pump, microvac dust sample from the back of a ceiling tile.

Figure 9. Collecting a microvac dust sample from the back of a ceiling tile.

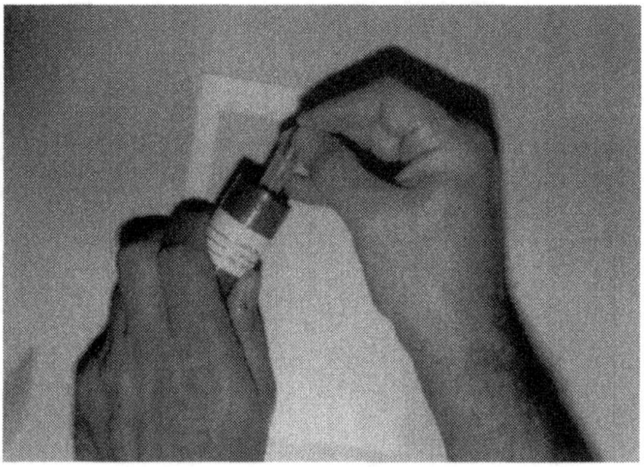

Figure 10. Capping the microvac sampler.

Collection Procedure

1. Mark 100 square centimeters of sampling area or use a template. For areas other than 100 square centimeters, measure the length and width of the surface and record.

2. The area is vacuumed by moving the filter cassette nozzle across the area until there is no visible dust remaining in the area, or, in the case of no visible dust present at the beginning of sampling, sample for 2 minutes. The nozzle should contact the surface.

3. At the end of the sampling, the cassette is turned upward before shutting off the power to the pump.

4. The nozzle is plugged and the cassette end sealed with an end plug.

5. Mark the sample number, location, and date on a plastic Ziploc™ bag and place the microvac sample in the bag. Make sure that the assembly is appropriately packaged to prevent separation of the nozzle from the cassette and sealed in a Ziploc™ plastic bag. Place an unused microvac sampler in a bag and mark as a field blank. Ship to the laboratory for analysis. Specify TEM analysis.

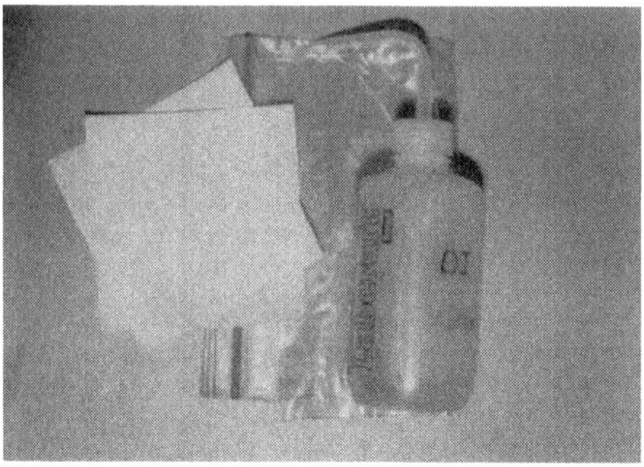

Figure 11. Particle-free cloth, prefiltered distilled water and plastic bag for wet wipe sampling. The cloth is cut in a square of 100 cm^2.

Figure 12. Placing the moist cloth over the area to be sampled.

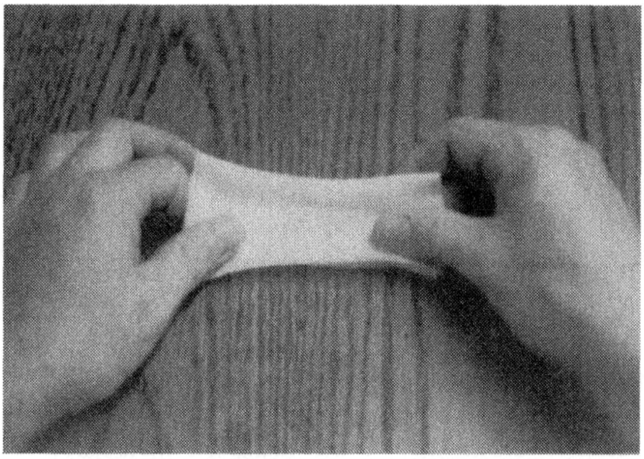

Figure 13. Fold the cloth and rewipe the area.

WIPE SAMPLING

Intended Use

A collection procedure for determining the presence and type of asbestos. In some cases, it can be used to determine a quantitative value for the amount of asbestos present.

Materials

Tex-Wipe™ Alpha 10 cloths
Ziploc™ plastic bags or other tight-fitting containers
Sample labels

Collection Procedure

1. Mark 100 square centimeters of sampling area or use a template. For areas other than 100 square centimeters, measure the length and width of the surface and record.
2. Moisten, prior to sampling, Tex-Wipe™ brand cloths with distilled water. The towel should not be dripping.

3. Using an "S" type wiping motion, wipe the sampling area. Fold the cloth over to expose a clean side of the cloth and wipe the sampling area again in a direction 90° from the first wipe.

4. Mark the sample number, location, and date on a plastic Ziploc™ bag and place the sample cloth in the bag. Place an unused but moistened cloth in a bag and mark as a field blank. Ship to the laboratory for analysis. Specify TEM analysis.

CARPET - COLLECTION FOR ULTRASONIC SHAKING

Intended Use

A procedure for determining the presence and type of asbestos in carpeting or textile material.

Materials

A razor or utility knife
Tight-fitting plastic bottle or box or Ziploc™ plastic bag
Sample labels

Collection Procedure

1. Mark 100 square centimeters of sampling area or use a template. For areas other than 100 square centimeters, measure the length and width of the surface and record.

2. Cut a piece (usually 10 centimeters by 10 centimeters) from the carpet with a razor blade or utility knife.

3. Place the carpet piece in a wide-mouth plastic jar or Ziploc™ bag.

4. Mark the sample number, location, and date on the sample container. Ship to the laboratory for analysis. Specify TEM analysis.

PASSIVE MONITORING TECHNIQUE

Intended Use

A procedure for determining the presence, type, and quantity of asbestos accumulating over a known time period.

Figure 14. Cutting a piece of carpeting for the bulk carpet sonication dust removal procedure.

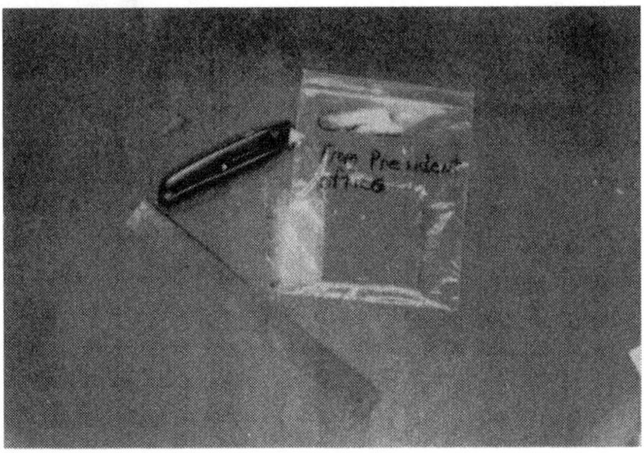

Figure 15. Place the carpet piece in a plastic bag or tight fitting container.

Materials

Forensic evidence tin or other metal or plastic shallow dish with a tightly fitting lid.

Indelible pen for marking on containers.

Collection Procedure

1. Mark the sample number and date on the sample container.
2. Place the tin in an area where it will not be disturbed and remove lid. Store lid.
3. Leave tin in place for a known period of time, usually 1 week to 2 years.
4. Open another tin for 15 seconds in the area where the test tin is to be placed. Mark as field blank.
5. Maintain a log of all pertinent sampling information, including sampling locations, dates, and times of sample placement and retrieval. Ship to the laboratory for analysis. Specify TEM analysis.

CHAPTER 5

Laboratory Analysis Procedures

Laboratory analysis procedures have been developed for a variety of sample collection media. The laboratory procedures described herein are those currently being used by a laboratory routinely involved with dust analysis.

The information that follows is general in nature. It is intended to assist experienced laboratory personnel and is not intended to be a specific, stepwise protocol. Procedures may change over time, and the reader is encouraged to consult with the EPA, OSHA, ASTM, or other groups involved with publishing protocols to obtain the most current versions. Laboratory cleanliness and the health and safety of laboratory personnel are not generally addressed, and the reader is cautioned that all procedures should be done according to good laboratory practices and in compliance with a properly prepared health and safety program.

CELLOPHANE TAPE—PLM ANALYSIS

In the laboratory, each piece of tape is divided into two pieces. One piece is archived. The other piece is divided into two parts. One part is mounted, adhesive side up, on a new glass slide, using a drop of 1.550 high dispersion (HD) refractive index oil underneath the tape and one drop on top of the tape under the cover slip. The same procedure is followed for the second piece of tape, using 1.690 or 1.700 HD refractive index oil. The slides are analyzed using PLM in which the asbestos fibers are identified using dispersion staining techniques.[1] It should be noted that the tape itself has a polarizing effect which sometimes makes the optical properties harder to view. In most cases, the dispersion staining can be observed but the colors may indicate a slightly higher refractive index.[2] Amosite asbestos will normally display vivid central stop dispersion staining colors in a liquid having a refractive index of about 1.680. However, when these fibers are on the tape, a liquid having a value of 1.690 or 1.700 should be used and the diagnostic colors for amosite will be observed. The approximate percentage of asbestos in the slide mount is estimated for each preparation and the two results are averaged. Cellophane tape samples are difficult to prepare for TEM analysis; therefore, this technique is not recommended.

ANALYSIS OF POST-IT™ NOTES BY SCANNING ELECTRON MICROSCOPY - ENERGY DISPERSION X-RAY SPECTROSCOPY (SEM-EDS)

In the laboratory, the Post-it™ notes are opened and examined under a stereomicroscope. If conductive tape is used, both adhesive surfaces are examined under low power (10–40×) to locate areas that may contain particles of interest. A portion of the adhesive surface(s) is cut out, using a new, single-edge razor blade. The excised section can be attached to a carbon planchet or attached to an aluminum stub, using carbon conductive tape. The sample is carbon coated and placed into the SEM-EDS. The excised area is examined at 150× for areas containing particles. Fibers and particles of interest may be photographed and an x-ray elemental spectrum collected. The types of particles or fibers found are noted on an analysis worksheet. The examination is considered qualitative. If no particles or fibers of interest can be located at 150×, the entire excised adhesive surface is examined at 600×. The use of the carbon tape or carbon sheet is optional but preferable to the Post-it™ note for some analyses because the Post-it™ note adhesive contains particles high in zinc and chloride that could interfere with the analysis. However, the carbon conductive tape or sheet may not release easily from some surfaces and may damage others, due to its greater adhesive strength. Large particles or debris may be mounted for SEM-EDS analysis by leaving them attached to the excised adhesive surface, or they may be removed from the adhesive on which they were collected and placed directly onto a carbon planchet for analysis.

MICROVAC SAMPLES

In the laboratory, the unopened microvac cassettes are wet-wiped and then prepared for analysis under clean room conditions. The cassettes and filters are thoroughly rinsed with a 50/50 mixture of alcohol and particle-free water and filtered through a 1 mm mesh sieve. The resulting solution is filtered through a second filter which is used in the analysis. The original collection filter is used during the sample collection procedure but otherwise is not used in the analysis.

Specifically, the plug from the nozzle of the sampler is removed and the cassette is filled with approximately 10 ml of a mixture of 50/50 alcohol and prefiltered water. The plug is replaced and the cassette is shaken vigorously by hand for 2 or 3 seconds. The entire cap of the cassette is removed and the suspension poured into a precleaned 200 ml glass medical specimen bottle. All visible traces of the sample are rinsed into the specimen bottle with a plastic squirt bottle of 50/50 solution. This procedure is repeated two additional times for a total of three washings. Next, the nozzle is rinsed 2 or 3 times into the specimen bottle. Typically, the total amount of 50/50 solution used in the rinse is approximately 70 to 75 ml. The suspension is then filtered through a 1 mm mesh sieve and collected in a specimen bottle. The liquid level in the specimen bottle is then carefully adjusted to 100 ml with prefiltered water. The pH of the water is adjusted to 3 or 4, using a 1.0% HCl or acetic acid solution. The sample container is capped and placed in a mild ultrasonic bath for 3 minutes to make a uniform suspension. After two minutes of settling, a measured volume of suspension is extracted with a graduated pipette inserted halfway into the sample solution. The aliquot is mixed with particle-free water in the filter funnel and filtered through a 0.22 µm pore size MCE filter backed by a 5.0 µm pore size filter. The 0.22 µm filter is dried and prepared according to the direct filter preparation procedure described in this chapter. A sample blank is prepared in an identical way as the sample.

PASSIVE DUST CANISTERS

In the laboratory, the outsides of the tins are wiped clean with a moistened cloth in a high-efficiency particulate air (HEPA) hood. The initial specimen preparation takes place in a clean HEPA- filtered negative pressure hood to avoid possible contamination of the laboratory and personnel. The canisters are opened and rinsed out with 50 ml of prefiltered water into a filter funnel. Either a glass filtering unit or new disposable plastic filter funnel unit (either 25 or 47 mm diameter) can be used. If a glass filtering funnel is used, the unit must be cleaned between each sample filtered. If a disposable filter is used, the top filter supplied with the apparatus may not be of the required pore size and must be discarded. Assemble the unit with a properly sized neoprene or rubber stopper, and attach the funnel to the 1000 ml side arm vacuum flask. Place a 5.0 µm pore size MCE (backing filter) on the support pad. Wet it with a few ml of prefiltered water and place an MCE filter (less than or equal to 0.22 µm pore size) on top of the backing filter. Apply a vacuum, ensuring that the filters are centered and pulled flat without air bubbles. Any irregularities on the filter surface are cause to discard the filter. The flask is returned to atmospheric pressure. The entire contents of the tin are filtered. Before the filter is dried, the filtering unit is disassembled and the filter carefully removed with tweezers. The completed sample filter is placed particle side up onto a precleaned, labeled, disposable plastic petri dish or other similar container. The filter is dried and prepared according to the direct filter preparation procedure described in this chapter. A laboratory blank is prepared in an identical way as the sample, except that a tin which has not been opened in the field is used.

CARPET SAMPLES

In the laboratory, 5 centimeter by 5 centimeter squares of carpet are cut and placed carpet side down in a 1,000 milliliter beaker containing 100 milliliters of 0.1% solution of the surfactant aerosol OT or a 0.002% solution of the surfactant methyl cellulose in particle-free water. The beaker is placed in an ultrasonic bath for 30 minutes. The carpet piece is removed and rinsed into the beaker with 100 milliliters of particle-free water. The entire suspension (200 milliliters) is then shaken vigorously by hand to disperse the particles and then allowed to sit for two minutes to allow the denser particles to sink or the lighter particles to float to the top. At this time, three measured aliquots of different volumes (usually one, ten, and fifty milliliters) are extracted with disposable graduated pipettes 1/4 to 1/2 inch below the water surface in the beaker. The aliquot is mixed with particle-free water to make 50 milliliters and filtered through a 0.22 µm pore size mixed cellulose ester (MCE) filter. If by visual observation the initial beaker contains large non-asbestos particles (carpet material), the entire suspension is passed through a coarse stainless steel mesh screen before filtration through the 0.22 µm filter. The filters are dried and prepared according to the direct filter preparation procedure described in this chapter. At least two TEM grids from different areas of the filter are prepared for each aliquot. After the three filtrations are completed, the remaining suspension is transferred to a graduated cylinder and the volume recorded. This volume is added to the volumes of the measured aliquots to obtain the volume of the sample. This accounts for the variable amount of water absorbed in the carpet during processing. A sample blank is prepared in an identical way as the sample, although no carpet segment is actually used in the ultrasonic procedure.

DIRECT PREPARATION OF MCE FILTERS

Acetone fusing is a common procedure for rendering the surface of the filter flat. Using a 7 mm cork borer, remove a section from any quadrant of the sample or blank filters. Place the filter section particle side up on a clean microscope slide. Affix the filter section to the slide with a gummed page reinforcement or other suitable means. Label the slide with a glass scribing tool or permanent marker. Prepare fusing dish from a glass petri dish and a metal screen bridge. One way to prepare the fusing dish is to make a pad from 5 to 6 ashless paper filters and place them in the bottom of the petri dish. Place the screen bridge on top of the pad and saturate the filter pads with acetone. Because acetone is flammable, this procedure should be done in a fume hood. Place the slide on top of the bridge in a petri dish and cover the dish. Wait approximately five minutes for the sample filter to fuse and clear.

Another commonly used method of collapsing the filter is called the DMF-acetic acid method. Place a drop of clearing solution that consists of 35% dimethylformamide (DMF), 15% glacial acetic acid, and 50% ultrapure water (volume to volume) on a clean microscope slide. Gauge the amount used so that the clearing solution just saturates the filter. DMF is a toxic solvent and should only be used in a fume hood. Carefully lay the filter segment, sample surface upward, on top of the solution. Bring the filter and solution together at an angle of about 20° to help exclude air bubbles. Remove any excess clearing solution with filter paper. Place the slide in a vented oven or on a hot plate, in a fume hood, at 65–70°C for ten minutes.

In addition to the collapsing, plasma etching of the collapsed filter is necessary. The microscope slide to which the collapsed filter pieces are attached is placed in a plasma asher. Because plasma ashers vary greatly in their performance, both from unit to unit and between different positions in the asher barrel, it is difficult to specify the conditions that should be used. Insufficient etching will result in a failure to expose embedded fibers, and too much etching may result in the loss of particles from the filter surface. As an interim measure, it is recommended that the time for etching be calculated on the basis of the final observations of the sample in the transmission electron microscope. Place the glass slide containing the collapsed filters into a low-temperature plasma asher, and etch the filter. Carbon coating of the collapsed and etched filters is required. Carbon coating is performed with a high-vacuum coating unit, capable of less than 10^{-4} torr (0.013 Pascals) pressure. Units that are based on evaporation of carbon filaments in a vacuum generated only by an oily rotary pump have not been evaluated for this application and must not be used. Carbon rods used for evaporators should be sharpened with a carbon rod sharpener to necks of about 4 mm long and 1 mm in diameter. The rods are installed in the evaporator in such a manner that the points are approximately 100 to 120 mm from the surface of the microscope slide held in the rotating and tilting device. The glass slide holding the filters is placed on the rotation-tilting device, and the evaporator chamber is evacuated to a pressure of at least 0.013 Pa. The evaporation is performed in very short bursts, separated by three to four seconds, to allow the electrodes to cool.

An alternate method of evaporation is by using a slow continuous applied current. An experienced analyst can judge the thickness of the carbon film to be applied. Some tests should be made first on unused filters. If the carbon film is too thin, large particles will be lost from the TEM specimen, and there will be few complete and undamaged grid openings on the specimen. If the coating is too thick, it will lead to a TEM image that is lacking in contrast, and the ability to obtain electron diffraction patterns will be compromised. The carbon film should be as thin as possible and still remain intact on most of the grid openings of the TEM specimen.

To remove the last of the filter material and leave a grid with particles caught in the carbon film, the samples must be prepared in a Jaffe washer. There are several styles of Jaffe washers. One is made by placing several pieces of lens tissue on a stainless steel bridge. The pieces of lens tissue need to be large enough to completely hang over the bridge and into the acetone or DMF. In a fume hood, fill the petri dish with acetone (or DMF) to one quarter of the level in the dish.

The TEM grids are first placed shiny side up on a piece of lens tissue or filter paper so that individual grids can be easily picked up with tweezers. Two or three grids are prepared from each sample. Using a curved scalpel blade, excise three square (3 mm × 3 mm) pieces of the carbon-coated MCE filter from the glass slide. These squares are selected from random areas on the filter. The square filter piece is placed carbon side up on top of a TEM specimen grid. The whole assembly (filter/grid) is placed onto the saturated lens tissue in the Jaffe washer. The TEM grid sample filter preparations are placed on the same piece of lens tissue in the Jaffe washer. The lid is then placed on the Jaffe washer and the system is allowed to stand for several hours, preferably overnight. Alternately, place the grids on a low level (petri dish is only filled to the 1/8 mark) DMF Jaffe washer for 60 minutes. Then, to that washer, add enough solution of equal parts DMF/acetone to fill the washer up to the screen level. Remove the grids after 30 minutes if they have "cleared"; i.e., all filter material has been removed from the carbon film, as determined by inspection in the TEM. TEM grids used must have known grid opening area. Determination of this area can be done by measuring at least 20 grid openings in each of 20 random 200-mesh copper grids for a total of 400 grid openings for every 1000 grids used. This measurement is done by placing the 20 grids on a glass slide and examining them under the optical microscope. Grid area measurements can also be made at the TEM at a calibrated screen magnification of between 15,000 and 20,000×. Typically, one grid opening is measured for each grid examined. Grid openings are measured in both the x and y directions and the area is calculated. The use of precalibrated TEM grids is also acceptable.

REFERENCES

1. **Perkins, R.L. and Harvey, B.W.,** Test Method: Method for the determination of asbestos in bulk building materials, USEPA, Research Triangle Park, NC, EPA/600/R-93/116, July 1993.
2. **Mefford, D.C.**, Tape procedure for dust analysis, unpublished notes, DCM Science Laboratory, Golden, CO, 1989.

CHAPTER 6

Data: Levels of Asbestos in Dust

A considerable number of settled dust analyses for asbestos using TEM have been performed following the microvac sampling procedure. Some data also exist for analyses of carpet using the bulk carpet sonication technique. The values of asbestos in the dust have been usually given in terms of asbestos structures per square foot (s/ft^2) or asbestos structures per square centimeter (s/cm^2). To change from structures per ft^2 to per cm^2, divide the s/ft^2 by 929.

Typical asbestos concentration values in the surface dust as determined using the microvac technique were published for a variety of sites containing ACM.[1] A wide range was evident, from 6.5 s/cm^2 to over 4.3 million s/cm^2, averaging near 1000 s/cm^2 (one million s/ft^2). Typical values for carpet dust determined by both the microvac technique and bulk carpet sonification technique were also included. Typical carpet dust values for the microvac method ranged from 8.6 s/cm^2 to 2.2 million s/cm^2, averaging about 1000 s/cm^2. Bulk sample carpet sonification produced values starting at 32,000 s/cm^2 and ranging to 4.4 million s/cm^2, averaging about 108,000 s/cm^2. In studies comparing the microvac procedure to the bulk carpet sonification procedure, it has been found that the microvac apparently collects only 1–3% of the dust in the carpet.[2–3] Carpet appears to act as a reservoir for asbestos.[4]

Although limited, some settled dust data exist for samples taken from buildings which were constructed without ACM's. Microvac dust samples from three government buildings which did not have ACM's in the State of Maryland showed levels of asbestos fibers less than 200 s/cm^2. Surface samples collected in a western state in buildings with no ACM's were below the microvac technique usual detection limit of about 160 s/cm^2.[5]

As might be expected, areas that are normally cleaned on a routine basis are generally lower than areas where dust accumulates undisturbed over a period of time. In an office building with ACM's, samples were collected in a number of similar rooms in which the owner was able to document the last date that the cleaning service had cleaned that particular area. There was a general trend toward lower concentrations in the rooms that had been cleaned the most recently. The values ranged from 320 s/cm^2 down to 86 s/cm^2.[6]

The average level of asbestos in the settled dust which had accumulated on the floor, carpeting, and athletic mat of a gymnasium which had asbestos-containing acoustical plaster was 9,700 s/cm^2. The gym had been closed to use for about a year. A storage room in the same facility with similar asbestos-containing acoustical plaster had an area weighted mean level of asbestos in the surface dust of 870,000 s/cm^2.[7]

Information about the amount of asbestos in one well-used building carpet was collected from a cafeteria in the Social Security Administration Building in Baltimore, Maryland. The cafeteria had an acoustical plaster ceiling containing 1 to 5% chrysotile. All furnishings from the cafeteria and the carpet had been vacuumed with a conventional dry vacuum cleaner twice before samples were collected. Previous use and traffic patterns were not taken into account in collecting the samples. The samples were collected in a random manner and some samples may have been from areas where an appliance such as a soft drink machine may have stood previously. The average level of asbestos on the surface of the carpet as determined by the microvac technique was 40,000 s/cm^2. The average level of asbestos in the pile of the carpet as well as on the surface (as determined by the bulk-carpet sonication procedure) was 1.7–3.1 million s/cm^2.[8–9]

Microvac samples collected of dust resulting from a breech in containment of an asbestos abatement area into a store which had been cleaned the previous day showed levels of asbestos of about 100,000 s/cm^2.

Levels of asbestos in layers of dust collected by microvac from the tops of ceiling tiles and light fixtures below asbestos-containing fireproofing have in some cases been over 1 billion s/cm^2.

Based on the authors' experiences, levels of asbestos in settled dust as determined by the microvac technique are considered low if less than 1000 s/cm^2. Levels above 10,000 s/cm^2 are considered generally above background. Levels above 100,000 s/cm^2 are considered high and in the range of a significant accidental release from an abatement site.

Some data are available on the rate of accumulation of asbestos fibers on horizontal surfaces such as ceiling tiles below fireproofing. The monitoring was done using passive monitoring collectors modeled as described in Chapter 4. In one set of samples collected in a school building, passive monitoring tins collected after 14 days and over 70 days showed accumulation rates of about 200 asbestos s/cm^2/day. In another project, tins set out in a boiler room below fireproofing showed accumulation of about 600 s/cm^2/day.[10]

REFERENCES

1. **Wilmoth, R.C., Powers, T.J., and Millette, J.R.**, Observations on studies useful to asbestos O&M activities, *Microscope*, 39, 299, 1991.
2. **Millette, J.R., Clark, P.J., Brackett, K.A. and Wheeles, R.K.**, Methods for the analysis of carpet samples for asbestos, *Environ. Choices Tech. Supplement*, 1(2), 21, 1993.
3. **Kominsky, J.R., Freyberg, R.W., Chesson, J., Cain, W.C., Powers, T.J. and Wilmoth, R.C.**, Evaluation of two cleaning methods for the removal of asbestos fibers from carpet. *Am. Ind. Hyg. Assoc. J.*, 51, 9, 500, 1990.
4. **Wilmoth, Powers, and Millette.**
5. **Hays, S.M. and Millette, J.R.**, Decon: a case study in technology, *Asbestos Issues*, 42, February 1990.
6. **Wilmoth, Powers, and Millette.**
7. **Keyes, D.L., Chesson, J., Hays, S.M., Hatfield, R.L., Ewing, W.H., Longo, W.E. and Millette, J.R.**, Re-entrainment of asbestos from dust in a building with acoustical plaster, *Environ. Choices Techn. Suppl.*, 6, 1992.
8. **Millette, Clark, Brackett, and Wheeles.**
9. **Kominsky, J.R., Freyburg, R.W. and Brackett, K.A.**, Evaluation of three cleaning methods for removing asbestos from carpet - determination of airborne asbestos concentrations associated with each method, Report for USEPA Contract No. 68-CO-0016, 1991.
10. **Hays, S.M. and Millette, J.R.**, Use of observational data and experimental studies in developing better operations and maintenance plans, Presented at Environ. Management '92, Pittsburgh, PA, April 8, 1992.

CHAPTER 7

Information About the Source of Asbestos in Dust

One of the first scientific studies of settled dust occurred in 1833, during the voyage of the *Beagle* when Charles Darwin collected and analyzed settled dust that he found on board the ship while at sea.[1] The close examination of soils and related particles has been a fundamental activity of the forensic community for many years. Sherlock Holmes, the mythical detective of the Sir Arthur Conan Doyle books, published between 1887 and 1893, found it quite useful to study the soil and dust on a person to determine where they had been.[2] Locard, a real French detective, published several classic articles in the 1930's about the microscopic study of dusts to determine their origin.[3] *The Particle Atlas* also describes cases where the analysis of settled particles lead to a determination of the source of the particles.[4]

The value of studying settled dusts to gain information about previous airborne material has long been known in the field of outdoor air pollution studies. In 1963, Junge stated: "Although the composition of deposited material cannot be expected to be identical with that of the total suspended material, it may still reflect the major features with good approximation."[5]

Few settled dusts have been studied in detail to determine all their constituents. Arizona Road Dust is one which has been studied by MVA, Inc. and provides some information about what might be expected in a fine outdoor dust. Arizona Road Dust is a reference material used in a variety of tests as a standard representative outdoor dust. A sample of Arizona Road Dust as received from Phoenix, AZ, was analyzed at MVA, Inc. by light microscopy and computer controlled scanning electron microscopy (CCSEM) using a Philips 525 SEM. The analysis of the sample showed that most of the particles in this dust had diameters less than 2 micrometers. The mineral composition consisted mainly of the smectite clay, montmorillonite, with some quartz, mica, feldspar, and hornblende as shown in Table 1.

Studying some of the individual particles in the settled dust in which asbestos fibers are found can provide information about the original source(s) of the asbestos. Some of the particles are small replicas of the parent material, while others show some of the components found in the

Table 1. Mineral composition of Arizona road dust.

MINERAL	PERCENTAGE
Kaolinite	0.2%
Montmorillonite	28.1%
Silica (Quartz)	15.2%
Mica	14.8%
Alkali Feldspar	14.0%
Plagioclase Feldspar	12.7%
Hornblende	10.4%
Ilmenite	0.7%
Unidentified	3.4%

original in-place material. In some cases, the type of asbestos in the dust may give strong evidence of its source. Amosite and crocidolite, two of the amphibole asbestos forms, are almost never found in surface dust accumulations in a building except in the vicinity of an ACM having one of these forms. Their occurrence in the dust is strong evidence of fiber release from the local AC product. Because the sources of chrysotile are usually more numerous in a building, source identification requires an examination of the matrix particles to which the chrysotile fibers are attached. Knowledge of the location of ACM's in relation to the settled dust is also helpful in learning about the sources of the asbestos in the dust. If dust samples are collected from a surface below an asbestos-containing vermiculite/chrysotile fireproofing, the finding of matrices of vermiculite and chrysotile in the TEM while analyzing the sample suggests that some portion of the asbestos fibers making up the dust were originally from the fireproofing above where the sample was collected. The discovery of tremolite in the dust is a further indication that the dust came from the vermiculite/chrysotile fireproofing because tremolite is found as a minor contaminant in some vermiculites.

The sizes of the asbestos fibers in a dust sample as determined by a microscopic examination of the material will provide some information about the fibers' probable source. Generally, large asbestos fibers (greater than 5 µm in length) do not travel great distances from a source which is losing asbestos fibers by aging or deterioration; therefore, the discovery of large asbestos fibers in a dust sample suggests a local source of the fibers.

The analysis of samples of dust particles collected by using Post-it™ notes or other adhesive samplers has been found useful in studying the components of matrix structures.

The following photographs illustrate the use of Post-it™ note sampling. Figures 1–6 are examples of asbestos-containing particles that suggest a source of the asbestos in the dust. The scanning electron micrographs were taken using a JEOL Model 6400 SEM equipped with a NORAN Voyager x-ray analysis system.

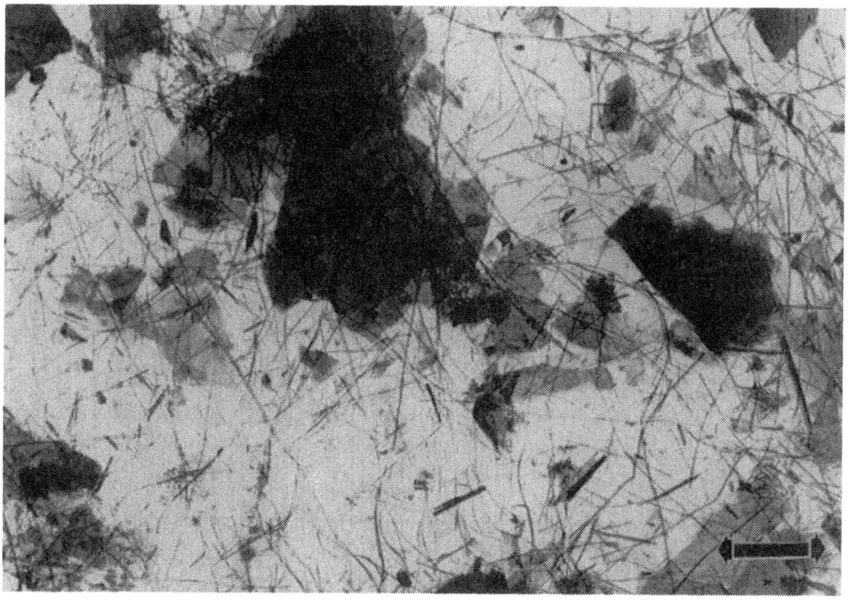

Figure 1. TEM micrograph from a microvac dust sample preparation which was too heavy for analysis. This micrograph shows an abundance of chrysotile asbestos associated with vermiculite. Some tremolite is also visible. Bar equals 5 μm.

Figure 2. SEM micrograph of a particle collected on a Post-it™ sampler. The particle contained mineral wool (m), chrysotile(c) and portland cement(p). Bar equals 10 μm.

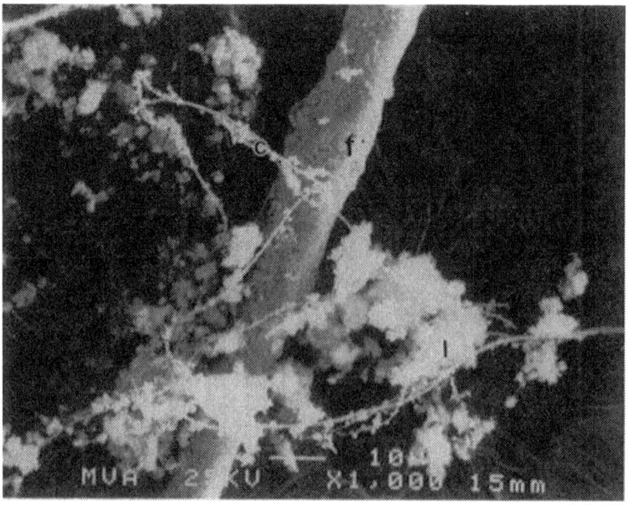

Figure 3. SEM micrograph of a particle collected on a Post-it™ sampler. The particle contained fiberglass (f), chrysotile (c), and lime (l). Bar equals 10 μm.

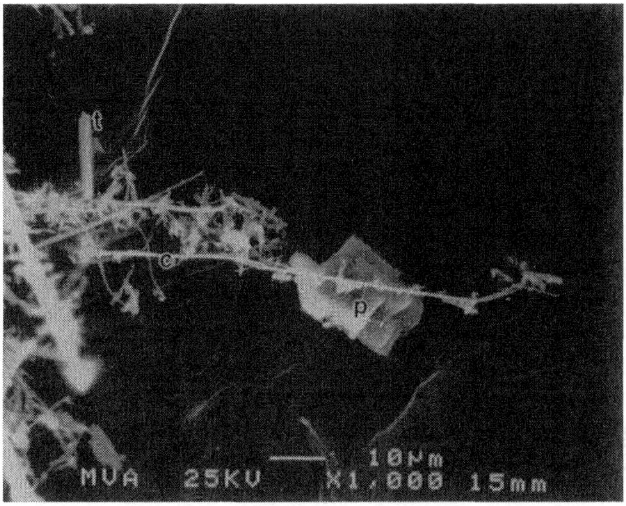

Figure 4. SEM micrograph of a particle collected on a Post-it™ sampler. The particle contained a perlite shard (p), chrysotile (c), and tremolitic talc (t). Bar equals 10 µm.

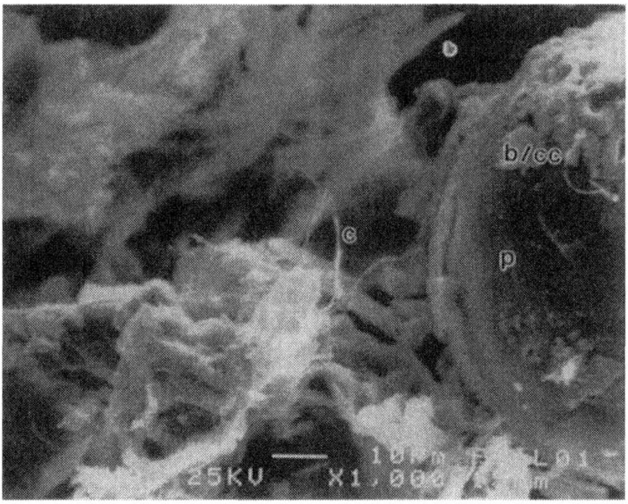

Figure 5. SEM micrograph of a particle collected on a Post-it™ sampler. The particle contained chrysotile (c), perlite (p), and bentonite (b) / calcium carbonate (cc) binder. Bar equals 10 µm.

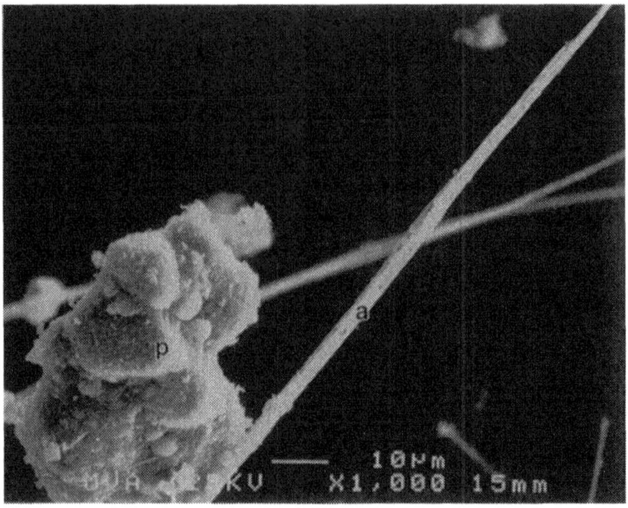

Figure 6. SEM micrograph of a particle collected on a Post-it™ sampler. The particle contained amosite (a) and portland cement (c). Bar equals 10 µm.

REFERENCES

1. **Darwin, C.**, An account of the fine dust which often falls on vessels in the Atlantic Ocean, *Q. J. Geol. Soc. Lond.*, 2, 26, 1846.
2. **Doyle, A.C.**, *The Complete Sherlock Holmes*, Barnes & Noble, New York 1122.
3. **Locard, E.**, The analysis of dust traces. *Amer. Jour. Police Sci.*, 1, 3, 276, 1930.
4. **McCrone, W.C., and Delly, J.G.**, *The Particle Atlas*, 2nd Ed, Ann Arbor Science Publishers, Inc., Ann Arbor, MI, 1973, 231.
5. **Junge, C.E.**, *Air Chemistry and Radioactivity*, Academic Press, New York, 1963.

CHAPTER 8

Resuspension of Settled Dust

The basic principles governing the resuspension of small particles in dust are complex and not well understood, in part because most dusts are a complex mixture of particles. The electrostatic forces between particles and other particles and between particles and surfaces are proportional to surface area, and the effect of the force is proportional to the mass of the particles. Smaller particles are influenced by these attractive forces more strongly than larger particles.[1] Individual small particles can be strongly attached to a surface by electrostatic forces. From work with nuclear materials, it was observed that small particles (below 2 µm) require local airflow velocities near 100 miles per hour to overcome surface attraction forces.[2] However, Bagnold, in his 1960 paper on the re-entrainment of settled dusts, reported that although very high threshold velocities are required to erode fine powders and settled dust, these particles can readily be ejected into the airflow by ballistic impacts of saltating larger grains.[3] Therefore, a single layer of very small particles, uniform in size, will be hard to dislodge from a surface, but mixtures of particles of different sizes may be more readily entrained. This is in agreement with the paper by Royster and Fish on techniques for assessing removable surface contamination of radioactive dusts which found that a mixture of two particle sizes gives a removal curve more nearly like the curve for the larger particle size.[4]

The study of how much asbestos is put into the air when settled asbestos fibers are disturbed is important because the resuspension of settled fibers is one of the main sources of exposure.[5–8]

Although limited, some data do exist of the analysis of air samples related to the disturbance of settled dust. Primarily the data consist of PCM fiber counts, and most of these reports of monitored activities do not include estimates of the amount of asbestos in the settled dust. Some of the data can be found in a report compiled by the Health Effects Institute (HEI)—Asbestos Research Literature Review Panel.[9]

In the 1971 study by Lumley, et. al., the resuspension of settled dust was shown to raise levels of fibers in the air.[10] With hardly any activity in storehouses insulated with crocidolite asbestos, the levels of fibers measured by PCM ranged from 0.01 to 1.26 f/cm^3. With 50 employees handling stored material, the levels ranged from 0.03 to 10.31 f/cm^3. When fallen asbestos material was disturbed on boxes or on the floor, the levels ranged up to 52.5 f/cm^3. More recent data obtained using TEM analysis showed levels of 30 s/cm^3 for activities disturbing settled dust originating from the fireproofing in a warehouse that included driving a forklift.[11]

In the Yale University Art and Architecture Building where chrysotile asbestos-containing material had contributed asbestos fibers to the dust on top of books, the fiber concentration during quiet conditions was found to be 0.02 f/cm^3.[12] During dry custodial sweeping and dusting, the levels increased to 1.63 and 4.02 f/cm^3 respectively. The levels proximal to the cleaning were 0.22 f/cm^3. Lifting of ceiling tiles containing debris on top caused levels of airborne asbestos averaging 5.45 f/cm^3 (PCM) in one study and ranged from 0.02 to 1.4 f/cm^3 (PCM) in another.[13–14] In the same studies, workers involved in inspecting or adjusting equipment above the ceiling tile produced levels of airborne asbestos averaging 1.3 f/cm^3 (PCM) in one situation and ranging from 0.04 to 0.9 f/cm^3 (PCM) in another.

Installing cable above ceiling tiles involves disturbing asbestos-containing dust and possibly some incidental contact with fireproofing. Data from two simulations performed in one school building show that this type of activity results in airborne levels of asbestos averaging 0.13 and 0.34 f/cm^3 (PCM) and 27 and 67 asbestos structures per cm^3 using TEM-Indirect.[15] A similar study performed in an office building with a different type of fireproofing showed average area samples collected during the cable installation of 3.5 s/cm^3 and personal samples averaging 26 s/cm^3 (TEM-Indirect).[16] Dry sweeping of the fireproofing dust after the cable pulling exercise showed airborne levels of 54 s/cm^3 in the general area sampling and personal samples averaging 61 s/cm^3 (TEM-Indirect).[16] Carpet cleaning of an artificially contaminated carpet using regular dry vacuuming methods produced airborne asbestos concentrations of 0.09 s/cm^3 (TEM-Direct).[17]

Re-entrainment of asbestos dust involves a redispersion or resuspension of fibers resulting from a variety of activities. Relating the airborne concentration of a material re-entrained from a surface and the surface dust concentration has been the subject of discussion for a number of years. The extreme case approach is to regard a room as a tightly sealed box into which all the asbestos fibers that are present in the settled dust are instantaneously released into the air. The total number of fibers present is divided by the volume of air in the room. In 1989, Guth presented an example of this in which a room 10 feet wide by 12 feet long by 12 feet high contained 0.01 f/cm^3 (PCM) in the ambient air and an average of 20 s/cm^2 (PLM) on the floor. He calculated that the maximum air concentration which would occur under the worst case conditions of total dust entrainment was 0.092 s/cm^3

or 9 times the ambient level.[19] Beard's presentation of a hypothetical room in which settled dust is raised is another example of this type of calculation. In his "room", one gram of dust containing 0.1% of asbestos has settled on each square meter of surface in the room which has an eight foot ceiling. He calculated that for fibers of a size 1 μm in diameter by 15 μm in length, the surface concentration of asbestos would be 3260 f/cm^2. If all of these fibers were made airborne, the fiber concentration in the air would be 13.4 f/cm^3. For fibers of a size 0.1 μm in diameter by 5 μm in length, the surface concentration would be 979,400 f/cm^2 and if all were made airborne, the concentration would be 4017 f/cm^3. Mr. Beard also calculated airborne concentrations where an activity of less intensity might result in only 1 of 1000 surface fibers being re-entrained and the situation where an activity of even less intensity might result in only 1 of each million surface fibers being re-entrained. For the 1 of 1000 case, the airborne concentration is reduced by a thousand fold. For fibers of a size 1 μm in diameter by 15 μm in length and a surface concentration of 3260 f/cm^2, the airborne concentration would be 0.0134 f/cm^3. A similar calculation can be done with the 1 of 1 million hypothetical case giving a value of 0.0000134 f/cm^3. Values for the 0.1 μm by 5 μm fibers which were present on the surface at a concentration of 979,400 f/cm^2 would be 4 f/cm^3 and 0.004 f/cm^3 respectively.[20]

Relating the airborne concentration of a material re-entrained from a surface and the surface dust concentration has been the focus of study for a number of different materials in addition to asbestos. Since the first work on a redispersion or resuspension factor (K) involving fission products by Chamberlain and Stanbury in 1951, K-factors have been determined for specific activities for a variety of radioactive compounds, beryllium, micro-organisms, metal oxide powders, and chrysotile.[21–27] Many of these studies have been reviewed by Sansone.[28] Mathematically, the K-factor is expressed as A/S, where A is the airborne concentration and S is the surface concentration of the material of interest. In controlled studies, the K-factor for various cleaning activities involving asbestos-containing dust is in the range of 10^{-5} to 10^{-6} cm^{-1}. Multiplying the settled dust asbestos concentration (structures per cm^2) by the K-factor (cm^{-1}) gives an estimate of the predicted airborne value in structures per cm^3 during a certain activity. Dr. Speight of the Radiological and Safety Division of the United Kingdom Atomic Energy Agency, Winfrith, Dorchester, England, used the concept of a re-entrainment factor to propose a way to calculate an acceptable level of crocidolite in the settled dust in 1983. He reasoned that "clearly, loose dust of any nature can become resuspended in air, and there must be a level of

Table 1. Examples of resuspension (K) factors.[28]

CONTAMINANT	CONDITIONS	K FACTOR (CM^{-1})	REFERENCE
^{131}Iodine-labeled dust	Active work in confined space	4.3×10^{-5}	21
Beryllium	Warehouse inventory	2×10^{-2}	22
Alpha Emitters	2 walkers, contaminated clothing	4.9×10^{-4}	23
Uranium Particles	Fans on, continuous cart movement	1.45×10^{-4}	24
Chrysotile	Handling contaminated materials	$2.0\text{-}4.2 \times 10^{-3}$	25
Microorganisms	Air jet Moist mopping	1.2×10^{-3} 2.0×10^{-4}	26
Zinc Sulfur Powder	Vigorous work with sweeping	1.9×10^{-4}	27

surface contamination at which the concentration of resuspended fibers, in the particular case of asbestos, approaches the recommended limits for breathing air".[29]

Some examples of resuspension factors for a variety of materials as published by Sansone are given in Table 1. In general, the more energetic the activity, the greater the K-factor. The K-factors determined for some asbestos activities are given in Table 2. The data were derived from controlled simulations where the levels of asbestos in the settled dust and during the activity could be monitored well. For most of the studies the K-factor is in the 10^{-5} to 10^{-6} range. When a forklift was used in the warehouse, a higher value was determined. Although there are factors such as ventilation rates and room size which may influence the value of the K-factor, the data here suggest that levels of asbestos in the dust of less than 1000 s/cm^2 will

Table 2. Resuspension K-factors determined from controlled asbestos studies.

ACTIVITY	SETTLED DUST LEVEL (s/cm^2)	AIR LEVEL (s/cm^3)	K-FACTOR (cm^{-1})
Gym-Athletic Activities	9,700	0.23	2.4×10^{-5}
Cleaning Storage Area	870,000	2.7	3.1×10^{-6}
Warehouse (with fork-lift)	8,200	29.7	3.6×10^{-3}
Cable Pull	2,000,000	28.9	1.4×10^{-5}
Broom Sweeping	760,000	53.6	7.1×10^{-5}
Conventional Carpet Cleaning	23,000	0.09	3.9×10^{-6}

not give rise to significant levels of asbestos fibers in the air during disturbing activities unless some very energetic physical force such as a forklift is involved. General cleaning, even dry broom sweeping or vacuuming, will not tend to cause air levels over 0.01 s/cm^3 if the settled dust levels are less than 1000 s/cm^2.

REFERENCES

1. **Lloyd, P.J.**, Surface Forces. Lecture 27, Particle Characteristics Conference, Loughborough University of Technology, Sept. 1967.
2. **Wilmoth, Roger C., Powers, Thomas J., and Millette, J.R.**, Observations on studies useful to asbestos O&M activities, *Microscope*, 39, 299, 1991.
3. **Bagnold, R.A.**, The re-entrainment of settled dusts, *Int. J. Air Poll.*, 2, 357, 1960.
4. **Royster, G.W. and Fish B.R.**, Techniques for assessing "removable" surface contamination, *Surface Contamination*, B.R. Fish, Ed., Pergamon Press, Oxford, England, 299, 1967.
5. **Sebastien, P., Bignon, J., Gaudichet, A., and Bonnaud, G.**, Les pollutions atmospheriques urbaines par l'asbeste, *Rev. Franc Mal. Resp.* 2, 51, 1976.

6. **Sebastien, P., Villon-Gallard, M.A. Dufour, G., and Bignon, J.**, Measurement of asbestos air pollution inside buildings sprayed with asbestos, EPA-560/13-8-026, U.S. Environmental Protection Agency, Washington, DC, 1980.
7. **Guillemin, M.P., Madelaine, P., Litzistorf, G., Buffat, P., and Iselin, F.,** Asbestos in buildings, the difficulties of a reliable exposure assessment, *Aerosol Science and Tech,* 11, 221, 1989.
8. **Fodero, S.D.,** Removal and disposal of an environmental carcinogen: asbestos, *J. Environ. Health,* 40, 133, 1977.
9. **Health Effects Institute—Asbestos Research** (HEI-AR), Asbestos in Public and Commercial Buildings: A Literature Review and Synthesis of Current Knowledge, 1991.
10. **Lumley, K.P.S., Harries, P.G., and O'Kelly, F.J.**, Buildings insulated with sprayed asbestos: a potential hazard, *Ann. Occup. Hyg.*, 14, 255, 1971.
11. **Hays, S.M. and Millette, J.R.,** Use of observational data and experimental studies in developing better operations and maintenance plans, Environmental Management '92, Pittsburgh, April 8, 1992.
12. **Sawyer, R.N.,** Asbestos exposure in a Yale building: analysis and resolution, *Environ. Res.*, 13(1), 146, 1977.
13. **Hamilton, D.J.,** Macrophage stimulation and the inflammatory response to asbestos, *Envir. Health Persp.*, 34, 69, 1980 (as cited in ref 9).
14. **CONSAD Research Corp.,** Economic analysis of the proposed revisions to the OSHA asbestos standards for construction and general industry, OSHA J-9-F-0033, U.S. Department of Labor, Washington D.C., 1990 (as cited in ref 9).
15. **Keyes, D.L., Chesson, J., Ewing, W.M., Faas, J.C., Hatfield, R.L., Hays, S.M., Longo, W.E., and Millette, J.R.**, Exposure to airborne asbestos associated with simulated cable installation above a suspended ceiling, *Am. Ind. Hyg. Assoc. J.*, 52:11, 479, 1991.
16. **Ewing, W.M., Chesson, J., Dawson, T.A., Ewing, E.M., Hatfield, R.L., Hays, S.M., Keyes, D.L., Longo, W.E., Millette, J.R., and Spain, W.H.**, Asbestos exposure during and following cable installation, *Envir. Choices. Techn. Suppl.*, 1:2, 12, 1993.
17. **Kominski, J.R., Freyburg, R.W., Chesson, J., Cain, W.C., and Wilmoth, R.C.**, Evaluation of two cleaning methods for the removal of asbestos fibers from carpet, *Amer. Indus. Hyg. Asoc. J.*, 51, 9, 500, 1990.
18. **Health Effects Institute—Asbestos Research** (HEI-AR), Asbestos in Public and Commercial Buildings: A Literature Review and Synthesis of Current Knowledge, 1991.
19. **Guth, J.H.,** Interpretation of surface asbestos contamination data: the closed finite system model, Update No. 1., presented at the National Asbestos Council Technical Conference and Exposition, Phoenix, AZ, September 10, 1990.

20. **Beard, M.E.**, Raising the dust, presented at Settled Dust: Asbestos and Other Particulates, Georgia Tech Res. Inst., Atlanta, Georgia, Aug. 12, 1991.
21. **Chamberlain, A.C. and Stanbury, G.R.**, The hazard from inhaled fission products in rescue operations after an atomic bomb explosion, Atomic Energy Res. Est. Rept., HP/R 737, Harwell, UK, June, 1951.
22. **Hyatt, E.C., Schulte, H.F., Mitchell, R.N., and Tangman, E.P.**, Beryllium: hazard evaluation and control in research and development operations, A.M.A., *Arch. Indust. Health*, 19, 211, 1959.
23. **Tagg, B.**, in *Soc. for Rad. Protect.*, Internat. Symp. on the Radiological Protection of the Worker by the Design and Control of His Environment, Bournemouth, England, 18, April 1966 (as cited in ref 28).
24. **Glauberman, H., Bootman, W.R., and Breslin, A.J.**, *Surface Contamination*, Fish, B.R., Ed., Pergamon Press, NY, 169, 1970.
25. **Carter, R.F.**, The measurement of asbestos dust levels in a workshop environment, United Kingdom Energy Authority A.W.R.E. Report No. 028-70, Aldermaston, UK, 1970.
26. **Hambraeus, A., Bengtesson, S., and Laurell, G.**, Bacterial contamination in a modern operating suite 3: importance of floor contamination as a source of airborne bacteria, *J. Hyg. Cab.* 80, 169, 1978.
27. **Fish, B.R., Walker, R.L., Royster, G.W., and Thompson, J.L.**, *Surface Contamination*, Fish B.R., Ed., Pergamon Press, NY, 75, 1967.
28. **Sansone, E.B.**, Redispersion of indoor surface contamination and its implications, Mittal, K.L., Ed., in *Treatise on Clean Surface Technology*, Plenum Press, NY, 261–290, 1987.
29. **Speight, R.G.**, The evaluation of asbestos contamination on surfaces: a new approach with an old technique, *Microscope*, 31, 175, 1983.

CHAPTER 9

Using Settled Dust Data in Building Inspections and in Operations and Maintenance

DESCRIPTION OF BUILDING INSPECTION

The phrase "building inspection" can mean different things to professionals in different disciplines. In this context, building inspection describes the process of determining whether or not a building is constructed or equipped with materials that are suspected of containing asbestos, lead, or other hazardous materials. This inspection is done visually and it is conducted with varying degrees of thoroughness, depending upon the needs and wishes of the building owner. One type of visual inspection is simply a pedestrian tour of the building to determine if any suspect materials exist within the occupied spaces. A more thorough inspection would include the mechanical and interstitial spaces of the building, and the most thorough inspection might even include some investigative demolition of chase walls or dismantling of certain equipment to determine if hazardous materials are present.

Building inspections for hazardous materials which generate particles in settled dust have procedures similar to those used for asbestos in settled dust. The remainder of this chapter uses ACM's in buildings to illustrate these procedures.

In the typical inspection process, such as one required by the AHERA regulation for asbestos in schools, the inspector not only determines which building materials are possible ACM's but also looks for dust or debris which obviously originated from a suspected ACM. Results are noted in the inspection report. After the inspection is done and materials are identified that are suspected of containing asbestos, laboratory analysis by PLM of bulk samples of those materials must be done to determine if, in fact, the suspect materials do contain asbestos. The presence of dust or debris related to a positively identified asbestos-containing material can lead the inspector to conclude that such dust and debris should be cleaned up, using asbestos control techniques.

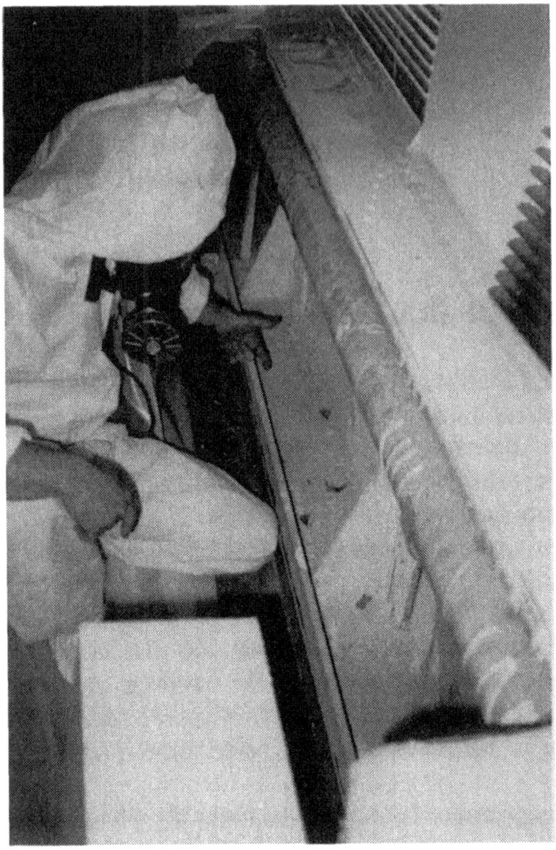

Figure 1. Dust and debris known to contain asbestos.

PURPOSE OF A BUILDING INSPECTION

A building inspection is intended to determine if ACM's exist in the building; if so, where they are located, what quantities of each type of material are involved, and what the conditions of the various materials are. "Condition" relates to whether the material is undamaged and in its intended state or whether damage and/or deterioration have/has occurred. This information is used to make some judgement about the relative exposure risk that the various materials present to the occupants of the building, the maintenance workers in the building, custodial personnel, vendors, contractors, etc. "Risk" is used here to mean the potential to release asbestos fibers into the air or to otherwise cause fibers to become airborne. The term is not used in the context of the continuing medical debate regarding morbidity of asbestos fibers of various sizes and mineral classifications nor in the context of the continuing scientific debate about threshold limit values. (These issues are discussed in the Health Effects Institute report *Asbestos in Public and Commercial Buildings: A Literature Review and Synthesis of Current Knowledge*[1] and the OSHA regulations 1910.1001[2] and 1926.58[3].) The theory is simple: You can properly deal with ACM only if you know its location, type, quantity, and condition.

An inspection report may then be used to develop an asbestos management plan. A comprehensive management plan should include a strategy and schedule for some form of abatement, such as removal and replacement with non-asbestos-containing materials, encapsulation, enclosure, or a strategy to manage the ACM's in place. This "in-place management" concept is referred to as an operations and maintenance (O&M) plan. (The phrase "operations and maintenance" is not unique to the asbestos control industry. Building owners routinely deal with operating and maintaining their facilities, asbestos notwithstanding, but the phrase has been adopted to mean procedures and work practices specifically designed to safely operate and maintain facilities with ACM's.) The inspection report is the first and fundamental step in a comprehensive approach to asbestos management and is used as a basis for future decisions about ACM's in the building and about the risk to occupants and other users of the building.

TRADITIONAL APPROACH TO BUILDING INSPECTION

The historic approach to building inspection involved primarily visual observation of the spaces within the building and PLM analysis of bulk samples taken from suspected materials to determine whether asbestos was actually present. A recognized approach to inspection can also be a visual observation only, with the owner then treating all suspect materials as if they actually contain asbestos. This eliminates the need for analytical costs at the time of the inspection. It may not be necessary to incur analytical costs until some point in the operation of the building when it becomes necessary to disturb a suspect material. At that time, analytical work can be done to determine if the suspect material contains asbestos. The traditional building inspection, therefore, is based primarily on non-instrumental means. Successfully performing the inspection requires that all relevant spaces are inspected and that visual access is gained to all nooks, crannies, and crevices of those spaces. It is often difficult to do a visual inspection of all parts of the building because some spaces are not accessible without dismantling parts of the building where ACM's may be concealed, such as the space between the interior wall and the exterior wall. To be comprehensively thorough, it is necessary to inspect these inaccessible areas or to note them in the inspection report as possible ACM or AC dust locations, to be inspected later if the need arises.

Proper inspection is, of course, directly related to the experience and skill of the inspector. The inspector must be very knowledgeable of building design and construction, familiar with what materials are considered suspect for asbestos, and adept at assessing the condition of those materials so that a determination of proper management protocol can be made. For example, if AC pipe insulation is so badly damaged that it has lost its cohesive integrity, as evidenced by dust and debris accumulation, the risk of exposure to airborne asbestos near the material is probably high. Some action should be taken to preclude the risk, and the inspector must be competent to judge whether the material can be successfully repaired or whether the damage is so severe that repair is impossible or economically impractical. If the latter is true, the inspector should recommend removal and replacement.

DUST SAMPLING AS AN ADDITIONAL COMPONENT OF INSPECTION AND OPERATIONS AND MAINTENANCE

The traditional building inspection can be enhanced by the use of settled dust sampling. This is particularly true for circumstances where the owner may desire to leave the ACM in place and institute a proper O&M plan so the building can be used with reduced risk of exposure to asbestos. Dust sampling can also be important data in the proper design of other abatement procedures such as removal, enclosure, or encapsulation. (See Chapter 10.)

Research simulating maintenance and custodial activities shows that the presence of AC dust can result in elevated levels of airborne fibers when the dust is improperly or inadvertently disturbed.

Two simulations of cable installation above a suspended ceiling were conducted by Keyes, et al. The structural steel above the ceiling was fireproofed with spray-applied ACM. Area air samples collected during the cable installations were 28.9 s/cm^3 and 100.2 s/cm^3.[4] (This paper is in Appendix: "Exposure to Airborne Asbestos Associated with Simulated Cable Installation Above a Suspended Ceiling.") A simulation of cable installation above a suspended ceiling was done by Ewing, et al., in a building fireproofed with spray-applied ACM, but with a product different from the AC product present in the previously mentioned simulations. Area air samples were collected during the cable installation and analyzed by TEM. The geometric mean was 3.5 s/cm^3. Area air samples were also collected during the cleaning activity following the installation. The geometric mean of the TEM data was .55 s/cm^3.[5] (This paper is in Appendix: "Asbestos Exposure During and Following Cable Installation.") A simulation of custodial cleaning (dry sweeping and dusting) in a room having a spray-applied AC acoustical plaster ceiling was conducted by Keyes, et al. Area air samples were collected during the cleaning and analyzed by TEM. The geometric mean was 3.0 s/cm^3.[6] (This paper is in Appendix: "Re-entrainment of Asbestos from Dust in a Building with Acoustical Plaster.")

O&M work practices, such as those published by the National Institute of Building Sciences,[7] are designed to avoid disturbance of ACM during the work activity or to properly control any intended disturbance. These control techniques focus on the AC building materials and are designed to reduce exposure risk to maintenance workers and building occupants. If, however, attention is given only to ACM's without regard to AC dust in the vicinity of the work activity, airborne asbestos concentrations can increase during the activity. (See Chapter 8.) Measurement of asbestos concentrations in settled dust and comparison of those concentrations to criteria for determining contamination, discussed in Chapter 6, can assist the O&M plan designer in determining whether special cleaning practices are necessary at the start of a maintenance activity. If asbestos in dust, settled on a surface which must be disturbed, is above ambient levels, the ACM's in the building are probably contributing to the surface loading. Depending on the surface characteristics (material, texture, etc.) and on the energy imparted to the surface during the disturbance, fibers may be released into the air. Proper cleaning of surfaces with high release potential prior to disturbance will reduce the risk of fiber liberation to the air. After surface cleaning, the maintenance activity (whether directly disturbing ACM's or not) can proceed as prescribed in the O&M plan. If dust measurements indicate that asbestos loading is not occurring in the settled dust, then the expense of special cleaning can be avoided.

If passive dust canister sampling has been done, then a rate of asbestos accumulation in settled dust can be determined. This sampling is especially useful for the top surfaces of ceiling tiles suspended beneath spray-applied AC fireproofing. As an example, it is common practice in commercial office buildings to replace interior finishes when tenant leases change. If interior partition walls are relocated, finish replacement may include all, or parts, of the ceiling for the affected space. If the accumulation rate of asbestos in settled dust is known for the building, the length of time required for the asbestos loading on the new ceiling tiles to reach levels of concern can be estimated. Until that time span has elapsed, the new ceiling tiles should not require special cleaning or handling, as long as visible ACM debris is not present on the tiles. Knowing the asbestos accumulation rate for a building is useful for many other maintenance and renovation scenarios.

Custodial work tasks do not usually involve the intended disturbance of ACM's. (Exceptions exist, such as the abrasive cleaning of vinyl asbestos tile and sweeping of ACM debris.) A significant portion of routine building cleaning, however, addresses surfaces which can collect dust, e.g., floors, countertops, shelves, and ledges. Cleaning techniques can be modified to control dust suspension into the air if dust sampling and analysis indicate the need. This may be especially relevant to surfaces which are cleaned infrequently, such as the top surfaces of exposed ductwork, pipes, or pendant-mounted light fixtures.

Chapter 6 provides asbestos in settled dust data and suggests surface concentrations of asbestos which, in the experience of the authors, may be cause for concern. There are not sufficient data, however, to set absolute thresholds for surface concentrations that apply universally. Dust concentrations that result in elevated airborne concentrations of asbestos depend, to varying degrees, on the characteristics of the surfaces involved and on the energy imparted to those surfaces by various activities. Chapter 10 discusses in more detail site specific selection of dust thresholds. Chapter 6 is appropriate for guidance, absent site specific data.

Given these considerations, it is prudent, after the building inspection and bulk sample analyses are complete, to consider dust sampling on surfaces where AC dust may settle and which may require disturbance during maintenance and/or cleaning of the facility. The ACM's which frequently cause heavy asbestos loading in dust are spray-applied surfacing products such as fireproofing and acoustical treatments. Damage to thermal systems insulation (TSI) can also result in high asbestos concentrations in dust.

In summary, dust sampling can assist in determining which work practices are appropriate and the stringency of those work practices. In the example of non-asbestos-containing ceiling tiles suspended beneath ACM, dust sampling may reveal that the ceiling tile surfaces are not laden heavily with AC dust, and, therefore, special precautions are not needed when moving those ceiling tiles to gain access for maintenance work. This information would reduce the cost of an O&M program for an owner; on the other hand, dust sampling could determine that the ceiling tiles are heavily laden with AC dust. Proper precautions would prevent not only exposure to maintenance workers but also inadvertent contamination of carpet and furnishings beneath the ceiling tiles during maintenance activities.

COLLECTION OF DUST SAMPLES

Determination of Sampling Locations

Dust samples should be considered for surfaces in the proximity of ACM's that are likely to create asbestos-laden dust, such as spray-applied materials (even those in good condition) or other ACM's which, because of damage or age, no longer have good cohesive and/or adhesive properties. Gravity is consistent and predictable, so a good place to start is directly beneath the ACM. Consider how air currents may cause asbestos fibers to migrate from their release source to other parts of the building. In making estimates of the amount of asbestos in dust in an area of interest, it is generally recommended that at least three samples be taken. Asbestos fibers in dust were found in one building on tops of ceiling tiles where no ACM existed above, but because the tiles bounded the return air plenum serving the entire floor, it was concluded that the asbestos fibers had migrated from other building areas on the same floor that had AC spray-applied fireproofing in the return air plenum.

Objectives for Sampling

It is appropriate to establish sampling objectives before beginning. If the objective is to design a proper O&M program, there may be implications for dust sampling which are different from sampling for the design of complete removal and replacement of ACM's. How to use dust sampling in design of abatement solutions other than O&M is addressed in Chapter 10.

The primary objective of surface dust sampling is to determine whether contamination exists which poses exposure risks to building occupants, users, or workers who might disturb that dust. These data can help decide whether to decontaminate surfaces, to ensure that surfaces are not disturbed, to replace surfaces, or to take no special precautions. Another objective may be to establish a criterion for cleanliness by determining what ambient concentrations of asbestos in dust occur in facilities constructed without ACM's or by determining asbestos in dust concentrations from the "cleanest" parts of the building in question. (See Chapter 10 for more discussion.) Dust sampling can also be used to establish a dust loading profile for building surfaces and then to ascertain whether the profile changes with time. This information is useful as a way to monitor the effectiveness of the asbestos management plan.

Figure 2. Dust and debris suspect for asbestos content.

Architectural Finishes

The use of dust sampling to determine contamination and hence to judge whether potential risks exist is related not only to the specific ACM's in the facility but also to the characteristics of surfaces to be sampled.

Architectural finishes on floors and on certain portions of walls are routinely used and disturbed by occupants. The disturbance patterns depend

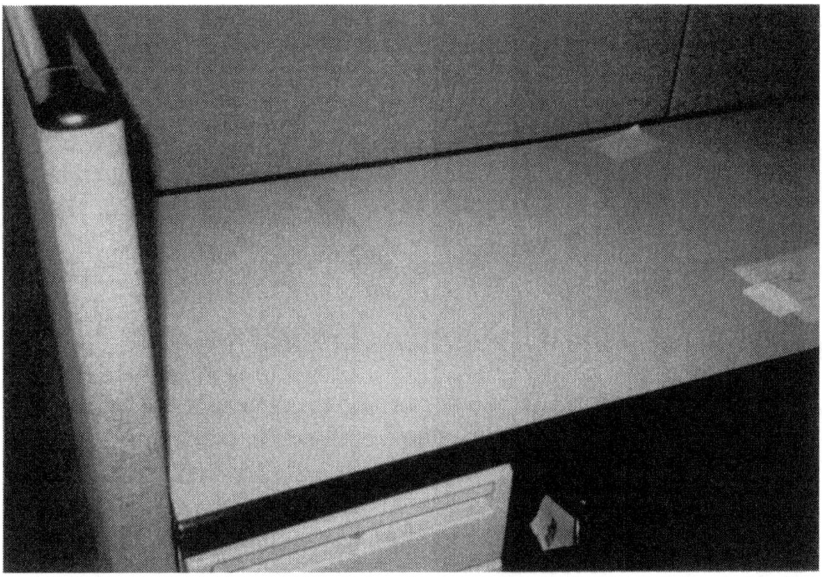

Figure 3. Dust suspect for asbestos content.

on the facility, its use, and the users. Custodial personnel usually clean floors frequently and may occasionally clean walls. In addition to architectural surfaces, furnishings should be considered, depending on where these furnishings are located and in what proximity to various ACM's. The type and texture of each surface is important. It is generally more difficult to decontaminate fabric surfaces or other soft, porous surfaces than it is hard, smooth surfaces. It is also very easy to disturb dust that is on a hard, smooth surface, while dust that is located on a fabric or other textured surface may require more aggressive disturbance to generate airborne fibers. K-factors, discussed in Chapter 8, may be useful in estimating airborne potential from surface sampling data.

The most direct way to correlate airborne asbestos concentrations resulting from a known asbestos loading for a given surface is a properly controlled and contained simulation of surface disturbance while collecting air samples. A relatively small amount of suitably gathered air concentration data, which can be related to surface loading, can be useful in determining where exposure risks exist from asbestos dust throughout an entire facility.

Predicting airborne concentrations from dust data must be done cautiously, however. Variables, such as surface characteristics, dust characteristics, work activities, differences among workers, and dust sample collection efficiencies, must be carefully considered when interpreting data. The data, when properly understood, are useful for O&M planning and O&M monitoring. Conclusions can be drawn before surfaces are disturbed, whereas air samples are collected during the disturbance and analytical results are available *after* the disturbance is initiated. Dust data must not be used, however, as a substitute for exposure monitoring of workers, which may be required by regulation or which may be required by good industrial hygiene practice.

Mechanical Equipment and Plenum Spaces

Mechanical equipment is another candidate for surface dust sampling. This includes the exterior of mechanical equipment which may need to be maintained from time to time and may include the interior surfaces of equipment or spaces that carry breathing air for building occupants. Return or supply air plenums with ACM's are prime candidates for dust sampling. Data indicate that settled dust inside such plenum spaces may not become airborne in the normal operation of the building mechanical system.[8] However, research clearly indicates that large concentrations of airborne fibers can be generated when these plenum surfaces are disturbed during maintenance or renovation.[9–10] Therefore, depending on the necessity of disturbing surfaces for maintenance and/or renovation, mechanical ductwork and/or plenum spaces should be considered for surface dust sampling.

METHODS OF COLLECTION

The method of collection for dust samples should be determined based on the surface being sampled and the objective of the sampling.

Microvac sampling, which has been described in Chapter 3, works very well for hard, smooth surfaces (which may be porous or nonporous) and can also work effectively on fabrics. Wipe samples, discussed in Chapter 3, may be more appropriate for smooth, metal surfaces such as the interior of unlined ductwork. Tape sampling, also described in Chapter 3, is generally used as a qualitative sampling technique to determine whether or not asbestos is present. It is generally not used as a quantification technique. Microvac and wipe sampling, however, can be done over a specified surface area,

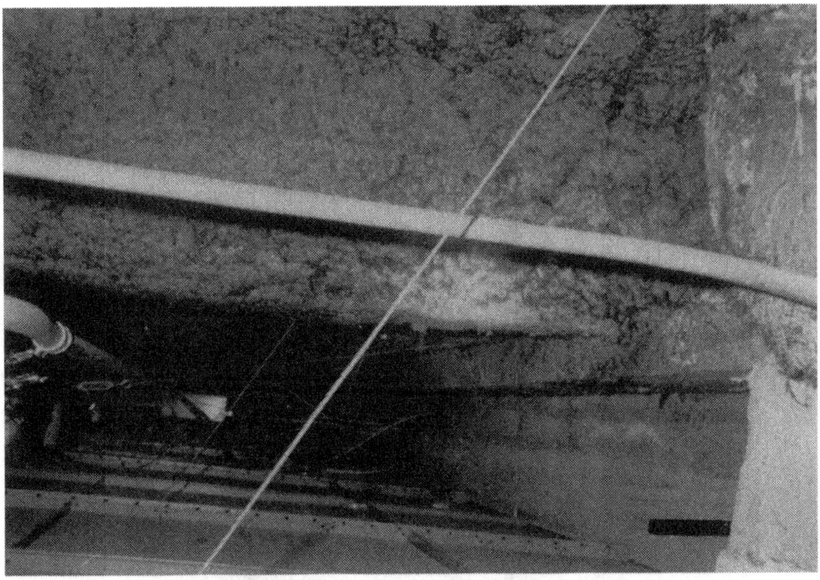

Figure 4. Plenum space that is a candidate for dust sampling.

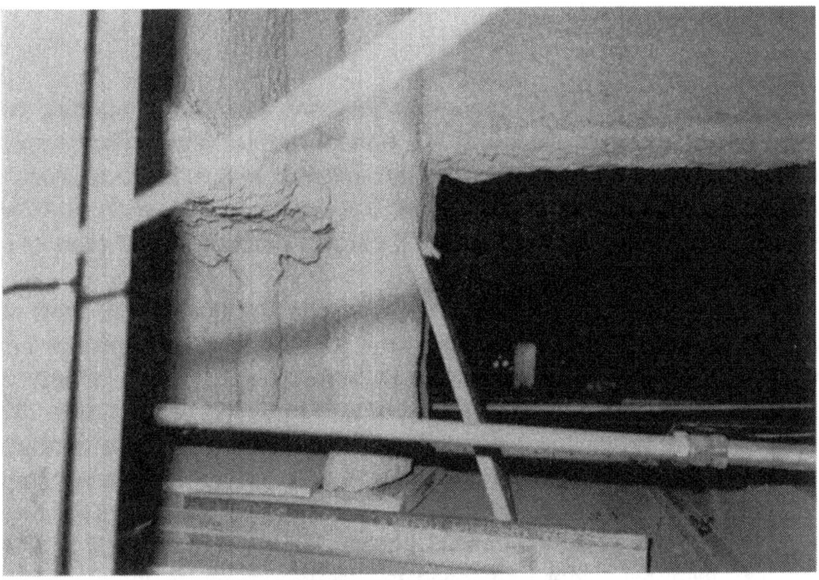

Figure 5. Plenum space that is a candidate for dust sampling.

Figure 6. Mechanical louvres that are candidates for dust sampling.

and then the analytical protocol can produce data that give dust concentrations in asbestos structures per unit of surface area. The sampling techniques that quantify the asbestos per unit area in dust are most useful to relate a measured surface loading to surface loading in areas without ACM's or to other standards that the owner may set for determining whether or not a surface is contaminated.

Forensic evidence tins are used to answer the question of how fast asbestos dust may be accumulating on a surface relative to one or more sources of release. This technique determines the number of asbestos structures settling on a given surface per unit time and per unit area. This information can be used to estimate how long it would take for a significant loading of asbestos to collect on top of new ceiling tiles, for instance, or on other new surfaces that may be installed in a building. This gives the owner an indication of appropriate cleaning cycles for surfaces, as well as where special procedures might be needed to safely perform custodial or maintenance activities.

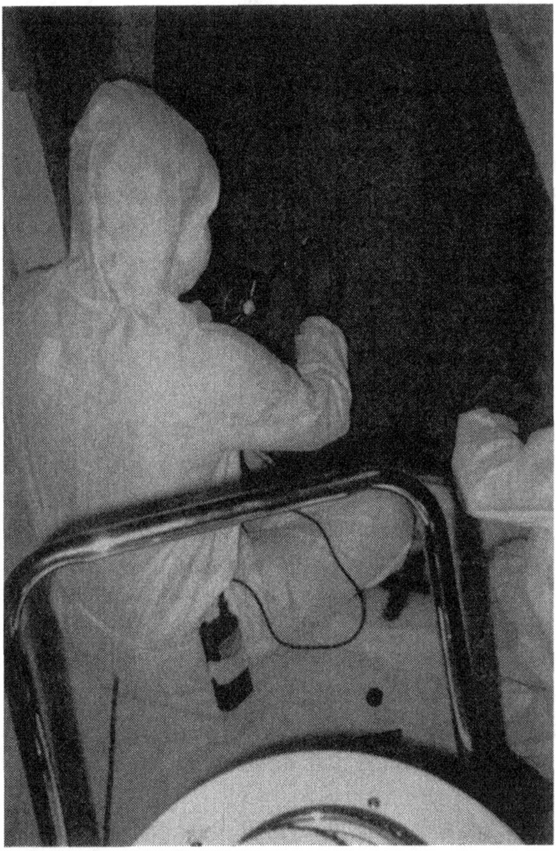

Figure 7. Microvac sample being taken - fabric.

An apparent contradiction exists; it is not unusual to find concentrations above ambient of asbestos in dust on tops of ceiling tiles in a return air plenum which contains AC spray-applied fireproofing. However, air sampling inside that plenum, even when samples are analyzed by TEM, may not detect elevated airborne concentrations. This apparent contradiction is explained in a paper entitled "Air and Dust Sampling in Return Air Plenums," contained in the appendices.[11]

INTERPRETATION OF DATA

Dust data should be interpreted either in relation to the collective experience which has been gained to date in this field (see Chapter 6) or should be related to ambient levels of asbestos in dust. Ambient levels are those levels found on surfaces in buildings with no ACM's and which are located in the same geographical locale in similar surroundings. If surface asbestos levels are above ambient levels, then a source, or sources, inside the building of concern may be releasing fibers which are accumulating on surfaces. Levels that are significantly higher than ambient probably indicate that special precautions may need to be taken. If the surface is to be decontaminated, an appropriate clearance level is probably the ambient level, as described above. In the long term, it is probably futile to clean surfaces to a level lower than the background for that geographic region, assuming similar surfaces and surroundings (urban, rural, etc.). The surface loading will eventually rise to ambient levels, but there may be short term justification, such as anticipated renovation, for cleaning some surfaces to below an ambient level, especially if ambient is above 10,000 s/cm^2.

Interpretation of dust data is often easier if a relationship between dust and air concentrations can be established. If the presence of asbestos in dust is confirmed on a given surface, the relevant question becomes whether activities that will happen to that surface will create an airborne concentration of asbestos which would pose an unacceptable exposure risk to occupants in the vicinity. If airborne asbestos data indicate that an exposure risk exists during surface disturbance, then the airborne data can be correlated to surface data to determine the appropriate surface clearance level. With sufficient data, extrapolation may be helpful to arrive at a surface clearance level.

Over time, as data are collected, it will be possible to construct curves with x and y axes of settled asbestos dust concentrations and airborne concentrations. Unique curves should exist for situations of similar surface characteristics and similar types of disturbances (e.g., low pile, commercial grade carpets vacuumed with a commercial, beater-bar style machine).

In conclusion, dust data are very useful to determine whether a surface is contaminated and to make decisions regarding whether to decontaminate the surface, to avoid disturbing the surface, or to remove and replace the surface with a clean surface. If the decision is made to decontaminate, dust data can also be used to determine when the decontamination process has been successful. Most importantly, proper collection of dust samples and proper analysis give a building owner information about where in a given facility stringent asbestos work practices may be necessary and where in a given facility such practices are not required.

IMPLICATIONS FOR THE ASBESTOS MANAGEMENT PLAN

Asbestos-in-dust data have implications for designing a management plan and for fine tuning an existing management plan to a given facility. Collecting this data allows the building manager to know more precisely what strategies are required, rather than making broad generalizations about asbestos management. Generalizations can be too lenient and, therefore, cause inadvertent airborne asbestos exposures in the building, or they can be too stringent and cause unnecessary expense to the building owner. In addition to inspection for suspect ACM's, prudent and proper collection of settled dust samples and proper analysis give the final piece of information which is needed to make appropriate decisions about asbestos management.

REFERENCES

1. **Health Effects Institute—Asbestos Research**, Asbestos in Public and Commercial Buildings: A Literature Review and Synthesis of Current Knowledge, Health Effects Institute-Asbestos Research, 1991.
2. **Occupational Safety and Health Administration,** 1910.1001, *General Industry,* OSHA Safety and Health Standards (29 CFR 1910), U.S. Dept. of Labor, OSHA 2206, June, 1981.
3. **Occupational Safety and Health Administration,** 1926.58, *Construction Industry,* OSHA Safety and Health Standards (29 CFR 1926/1910), U.S. Dept. of Labor, OSHA 2207, 1991.
4. **Keyes, D.L., Chesson, J., Ewing, W.M., Faas, J.C., Hatfield, R.L., Hays, S.M., Longo, W.E., Millette, J.R.**, Exposure to airborne asbestos associated with simulated cable installation above a suspended ceiling, *Am. Ind. Hyg. Assoc. J.*, 52, 11, 479-484, 1991.
5. **Ewing, W.M., Chesson, J., Dawson, T.A., Ewing, E.M., Hatfield, R.L., Hays, S.M., Keyes, D.L., Longo, W.E., Millette, J.R., Spain, W.H.**, Asbestos exposure during and following cable installation in the vicinity of fireproofing, *Environ. Choices Techn. Suppl.*, 12-18, March/April, 1993.
6. **Keyes, D.L., Chesson, J., Hays, S.M., Hatfield, R.L., Ewing, W.H., Longo, W.E., and Millette, J.R.**, Re-entrainment of asbestos from dust in a building with acoustical plaster, *Environ. Choices Tech. Suppl.*, 6, 1992.
7. **National Institute of Building Sciences**, *Guidance Manual: Asbestos Operations and Maintenance Work Practices*, NIBS Document Number 5076-7, National Institute of Building Sciences, Washington, DC, 1992.

8. **Hays, S.M.**, Air and dust sampling in return air plenums, presented at Environmental Management '92, The Ninth Conference and Exposition of NAC/The Environmental Information Association, Pittsburgh, April 8, 1992. (Included in appendices.)
9. **Keyes, D.L.,** et al.
10. **Ewing, W.M.**, et al.
11. **Hays, S.M.**

CHAPTER 10

Relevance of Settled Dust to the Design of Abatement Actions

INTRODUCTION

Abatement of most hazardous materials which generate particles in dust is similar in some aspects to abatement of ACM's. This chapter discusses the particulars of asbestos abatement projects as a means to illustrate these crucial considerations.

The use of settled dust measurements in building inspections for asbestos was discussed in the previous chapter, as was its use for in-place management of ACM's. In-place management, which is accomplished through the implementation of an O&M plan, includes the repair and removal of small quantities of ACM's, as dictated by the operational needs of the facility. In-place management is a recognized means of asbestos abatement. "Abate" means to lessen or reduce something, and in the case of asbestos, the term "abatement" means actions taken to lessen or reduce the risk associated with the mineral. The other familiar abatement techniques are encapsulation, enclosure, and removal and replacement. These terms are defined in the Glossary.

This chapter discusses the relation of settled dust measurement to encapsulation, enclosure, and removal projects. The reader is assumed to have a fundamental understanding of these techniques as presented in the EPA Purple Book.[1]

DUST CRITERION

At the time of this writing, no regulatory standards exist for asbestos in settled dust. Much experience has been gained by practitioners and much data have been collected over the past seven or eight years. This experience has led to development of sampling and analytical techniques, discussed in Chapters 3, 4, and 5, which are widely used and which produce valuable data. Data have produced conclusions about quantities of asbestos in dust that may create exposure risks in certain situations; these data are presented in Chapter 6. For dust sampling to be useful in a facility, the data from that facility must be compared to some criterion. The data in Chapter 6 is useful for comparison if a criterion of experience is deemed appropriate for a facility. If other criteria are desired or if data presented in this book were not collected under circumstances sufficiently similar to make valid comparisons, then other means may be legitimately used to develop a facility-specific criterion.

It is not unusual to find asbestos in dust in most urban settings, especially if one searches diligently enough with an electron microscope. The pertinent question is whether or not the surface concentration poses an exposure risk. The most direct way to answer this is to do a simulation, or series of simulations. The purpose is to perform, under tightly controlled and properly contained conditions, an activity which disturbs AC dust, such as vacuuming a carpet. The simulation should duplicate in all respects the activity of concern on the surface of concern. This is most easily done if the simulation area is small, and the area should be isolated, using appropriate asbestos abatement techniques, from the rest of the facility to prevent the spread of airborne fibers, if such are generated. The individuals performing the simulation must be protected with appropriate respirators and other personal protective equipment. If settled dust asbestos concentrations are known before the simulation and personal and area samples are collected during the simulation, then for that activity, settled dust data and airborne data can be correlated. With sufficient data sets, extrapolation may be appropriate to assist in estimating the surface concentration limit, below which that activity will probably not result in unacceptable airborne concentrations. This data may also assist in drawing useful conclusions about settled dust on *similar* surfaces and/or with *similar* activities. If the simulation results in airborne contamination, then asbestos decontamination procedures suitable for the level of contamination created by the simulation must be used to clean the air and the surfaces in the simulation area.

For this simulation approach to be useful in establishing a surface contamination criterion, the facility owner must have previously determined what airborne asbestos concentrations are acceptable. Applicable regulations dictate, of course, the least stringent airborne criterion which an owner must meet. Many building owners and managers opt for a lower air concentration limit than the regulations impose; e.g., some commercial office building owners use the AHERA clearance standard for schools, rather than the OSHA occupational standard, as the airborne limit for their facilities because the AHERA standard, which does not apply under federal law to office buildings, is more stringent than the OSHA standard. The selection of an airborne concentration criterion is discussed in the National Institute of

Building Sciences' *Guidance Manual: Asbestos Operations & Maintenance Work Practices.*[2]

Dust and air data correlation through simulation is probably not practical for small facilities, especially for landlords with low profit margins. The procedure is useful and practical, however, for large facilities where data can be generated, with a relatively small number of simulations, that is relevant to hundreds of thousands of square feet of space. Properly collected dust data can then be used to estimate, within reason, where exposure risks may exist.

The authors are knowledgeable of a variation on the use of activity simulation wherein an area with a known asbestos surface concentration is properly contained and the surfaces are disturbed using an aggressive air technique, such as that used for an AHERA abatement work area clearance. Air samples are collected and analyzed. This procedure, if done correctly, can be used to relate surface asbestos concentration to airborne concentration, but it does not have direct relevance to a particular work activity. It may be used to estimate what happens to airborne concentrations in a worst, or near worst, case scenario where all surface dust in the simulation area is disturbed coincidentally.

If historical data and/or simulations are not desirable to establish a dust criterion for asbestos, then dust data from other buildings may be used. Samples should be collected from buildings in the same locality which are used for similar purposes as the target building but which do not have ACM's, i.e., newer buildings.

Useful data may also be generated from the target building by sampling surfaces in areas of the building where there are no ACM's *and* where it is unlikely that asbestos fibers have migrated via air currents or otherwise. Sampling exterior surfaces may also be useful if the surfaces are those on which ambient airborne fibers might be expected to settle.

To avoid biased data when sampling surfaces, whether for target or comparison, consideration must be given to whether or not the surface has recently been cleaned, if it is an interior surface, or washed by a lawn sprinkler, rain, etc., if it is an exterior surface. Any of these procedures for setting criteria are likely to result in different criterion for different surface types, even within the same facility. This is because sample collection efficiencies vary with surface types, e.g., fabric versus polished wood, and because a given activity may produce different airborne concentrations from different surfaces with equal *measured* surface concentrations, e.g., sweeping smooth, polished wood versus rough, porous concrete.

A technique also exists to estimate airborne concentrations from surface concentrations by calculating a resuspension coefficient, called a "K" factor. This is discussed in Chapter 8.

ABATEMENT PROCEDURES

In the planning stage of an abatement project, the abatement designer must decide which abatement technique will be used to accomplish the owner's objectives for each of the ACM's to be addressed. Materials may be removed in their entirety and replaced, may be enclosed, or may be encapsulated. Large projects often involve different techniques for different materials; e.g., some acoustical AC plaster may be removed and replaced because it is in bad condition, some of the same material may be encapsulated because it is in good condition and not accessible to occupants, and all thermal systems insulation (TSI) may be removed and replaced to facilitate future maintenance. Once these conceptual decisions are made, the designer must then select the various means and methods by which these abatement ends are to be achieved.

Settled dust data can be useful in this design process. If an ACM is shown by microvac sampling or by passive monitor sampling to be contributing to surface loading over time, this may be a sufficiently compelling reason to remove and replace the material. Dust data can be used to determine if an ACM is releasing fibers which are migrating to other parts of the facility. This information might indicate that an enclosure is appropriate where, absent the migration, the material could have been managed in-place.

In one project, only one part of the building was fireproofed with a spray-applied ACM. This ACM existed inside an above-ceiling return air plenum, which was open to all other return air plenums in the building. No other friable ACM existed in the facility. Dust sampling and analysis indicated that asbestos fibers were being carried from the fireproofing by the air in the plenum to other parts of the building. The most cost effective solution was to isolate with permanent physical barriers the part of the plenum containing the fireproofing and to install below ceiling ductwork to route the return air around the isolated space.[3]

BOUNDARIES OF BUILDING DECONTAMINATION

Abatement projects are intended to address one, some, or all of the ACM's present in a facility. A well-designed abatement project should also consider whether parts of the building have been previously contaminated by deterioration of ACM's and/or past disturbance events (water leaks, maintenance activities, etc.). Building areas which will be part of a contained abatement work area will be thoroughly cleaned as part of the abatement decontamination and clearance procedures, so supplemental dust data

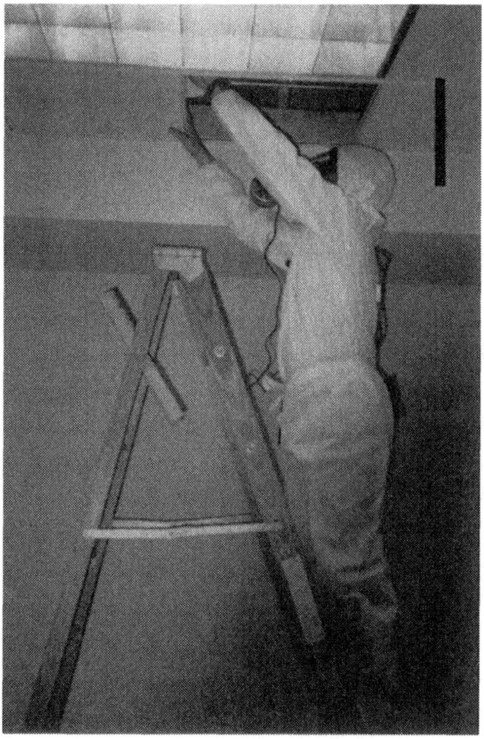

Figure 1. Moving ceiling tile which is contaminated with AC dust.

about contamination in those areas are not warranted, beyond that required to make the abatement technique decisions discussed earlier. However, dust sampling is useful in determining if building spaces outside the abatement work areas require any special cleaning in order to meet the owner's pre-determined cleanliness criterion (or criteria) for asbestos in settled dust. It is often most cost effective to do any necessary cleaning as part of the total abatement project.

DECONTAMINATION OF FURNITURE AND EQUIPMENT

Building furnishings and equipment can be treated in various ways during an abatement project. Many abatement techniques, especially removal, are best and most efficiently accomplished if these items are removed from the abatement work areas. This facilitates the abatement and protects the furnishings and equipment. Dust data, by comparison to the agreed cleanliness criterion, can establish which, if any, of these items need to be decontaminated, either before storage or before being returned to the abated and cleared spaces. Dust sampling and analysis can also determine if the cleaning techniques are successful.[4]

TREATMENT OF BUILDING SYSTEMS

Mechanical, plumbing, and electrical systems must be dealt with judiciously in an abatement project. Deciding how to treat external surfaces of such equipment is usually straightforward. Dust data are useful in determining if internal surfaces are contaminated and to what extent. A more informed decision can then be made regarding the cleaning of internal parts of the systems or replacing some parts. In one case, dust sampling led to the discovery that AC acoustical material had been inadvertently sprayed inside some duct work during building construction. This led to the decision to remove and replace certain sections of that ductwork.

WORK AREA CONTAINMENT

Settled dust sampling can also be useful in the design of abatement procedures for work area containment. In some circumstances, cleaning an area using asbestos control methods, such as HEPA vacuuming and wet wiping, prior to beginning an abatement action is required under the AHERA regulations for schools. If the building is not a school, whether or not cleaning the area prior to beginning the abatement process (referred to in the AHERA regulations as "precleaning") is a question that settled dust data can help answer. The presence of asbestos in dust can have implications on the establishment of containment for at least three reasons: 1) The activities associated with building the containment can create dust which can migrate to other parts of the facility. This is especially relevant if other parts of the building will not undergo abatement and the associated post-abatement cleaning and clearance and/or if the building will be partially occupied during abatement; 2) In some circumstances, containment can be established without contacting, or otherwise disturbing, the ACM which is to be abated. The presence of asbestos dust in the work area, however, can have implications for respiratory and other personal protection necessary for

Figure 2. Setting up containment.

workers who are preparing the containment; 3) Dust present in the work area prior to abatement may be more easily removed before containment is established than during post-abatement decontamination. Dust remaining on surfaces may be covered by critical barriers which remain in place during post-abatement cleaning and clearance. Removal of the critical barriers after clearance could expose an otherwise clean area to this pre-abatement dust. Dust data can assist in determining whether extensive precleaning is appropriate before full containment is established. It is appropriate, of course, to establish partial containment, or at least isolation, before the precleaning.

BREACH OF CONTAINMENT

If an abatement project is done improperly or carelessly, asbestos fibers can be spread outside the contained work area. It is common practice to do air monitoring outside the contained area as a means of detecting a breach in containment. If a breach is suspected or if air monitoring is inconclusive, dust sampling and analysis can assist in determining whether containment

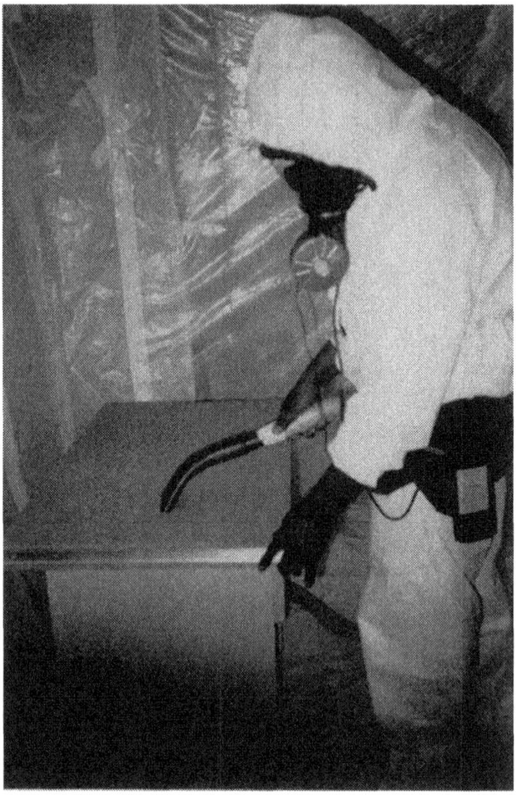

Figure 3. Decontaminating a surface - (HEPA vacuuming).

has been compromised and, if so, to what extent. If improper procedures are continuous and sustained, air sampling will probably detect movement of fibers outside the work area. If containment is compromised sporadically or episodically, air sampling may fail to detect the compromise. Assuming adequate pre-abatement dust data outside the work area, dust sampling can be used to monitor the containment integrity and to determine the seriousness of any known or suspected compromise.

If the compromise was an isolated event, air sampling outside the work area after the event might be effective only if it were done aggressively.

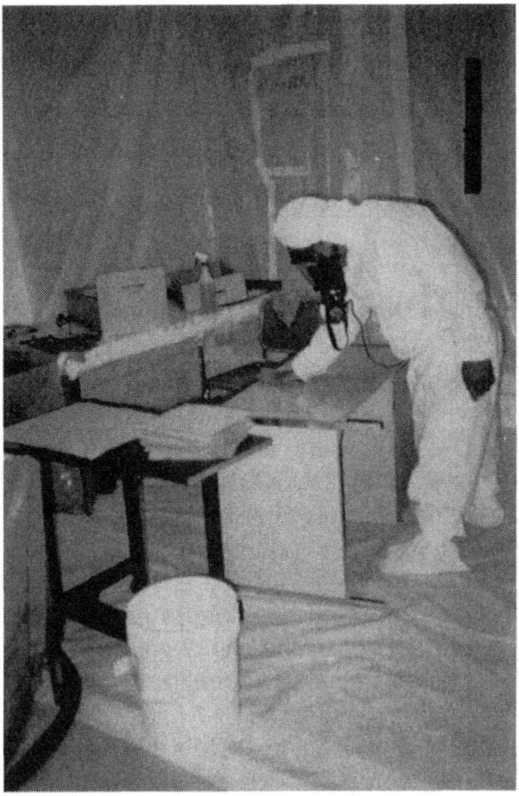

Figure 4. Decontaminating a surface - (Wet wiping).

This, of course, is not an appropriate air sampling strategy in uncontained areas. Dust sampling, therefore, has the advantage of providing useful data without creating additional risk outside the abatement containment.

One author has found asbestos in dust levels after a containment breach to be on the order of 100,000 s/cm^2. (See Chapter 6.)

DIAGNOSIS OF OTHER ABATEMENT ILLS

After an abatement area has been cleaned but has failed clearance testing, it is standard practice for the contractor to go back into the abatement area and clean again. This cleaning and sampling process continues until the area has been finally cleared. However, if clearance cannot be obtained, dust sampling may be used as a method to determine the source of the airborne fibers that are preventing clearance. In one instance, it happened repeatedly that clearance could not be achieved in an area, and dust sampling determined that the contractor was using HEPA vacuums contaminated with a layer of surface dust that was not visible to the naked eye. Once the HEPA vacuums were replaced with clean ones, clearance was obtained. Dust sampling in the work area before final air clearance sampling can also be useful supplemental clearance data.

DIAGNOSIS OF AN UNINTENDED BUILDING EVENT

Settled dust sampling is an extremely valuable analytical tool in answering contamination-related questions after an event, especially a catastrophic event in a building, such as a fire, explosion, or storm damage. Typically, in projects of this nature, after the catastrophic event has been controlled and the emergency response personnel have left the scene, the owner of the building is faced with the task of putting the building back in operation. To do that requires, at the very least, cleaning up the disturbances caused by the event, and if asbestos-containing materials have been disturbed either by the event or by the response to the event, the cleanup and reconstruction operation takes on a decontamination and perhaps an asbestos abatement component as well.

Fires very often cause damage to ACM's and cause asbestos fibers to be spread through the building in smoke and hot gases which move from the location of the fire to other parts of the facility. Storm damage may include water damage which can dissolve the matrices in which asbestos fibers are held and cause the asbestos fibers to be entrained in the water stream throughout the building. Once the water evaporates, the fibers dry out and can easily become airborne. If such distribution of fibers has occurred in the building, dust sampling is a quick and direct method of making that determination. Assessing the damage caused by earthquakes may have a settled dust component. It is well known that motion of a steel structure that has been fireproofed can cause delamination of the fireproofing. If delamination occurs in any significant amount, it can obviously be detected by visual inspection. However, the question remains: Did an earthquake which caused significant motion to the steel structure of the building cause the asbestos-containing fireproofing to increase its rate of dusting? That may not be obvious by simple visual inspection. If the rate of dusting has been determined in a building prior to such an event by the use of dust fall sampling,

Table 1. Settled asbestos dust concentrations in a high-rise building following a fire.

CONCENTRATION (million structures/ft^2)						
RANGE			MEAN		DATA POINTS	
Surface Type	Before Decon	After Decon	Before Decon	After Decon	Before Decon	After Decon
Misc. Horiz. Surfaces (tables, desks, cabinets, shelves, books)	<0.11-54	<0.13-8	3.6	0.37	30	22
Chairs (fabric)	<0.13-7.7	<0.09-2.2	1.5	0.33	19	29

then additional dust fall sampling after such an event can determine if that rate has increased or has remained stable.

Following are two case studies of settled dust sampling.

CASE STUDIES

Case Study #1

Data and other information supplied by the client to the industrial hygiene (I.H.) consultant related to asbestos-containing debris found inside a supply air duct. The supply air duct, of sheet metal construction, was connected to an air handling unit which served several rooms of a multi-floor building. Based on review of this data and information, the I.H. consultant provided an estimate of average airborne asbestos structure concentration which may have been achieved in the space served by the air-handling unit, due to the presence of debris in the duct.

Pertinent facts of the case, as understood by the I.H. consultant, were:

1. Debris was discovered inside a segment of a supply duct common to all spaces served by the air-handling unit.

2. Samples of this debris and samples of dust from inside the duct at a location downstream of the debris were positive for chrysotile asbestos by PLM analysis.

3. Analyses by PCM and TEM of ambient air samples taken in spaces served by the air-handling unit generally indicated airborne concentrations below detection limits.

4. Analysis by TEM of aggressive air samples taken from three rooms served by the air-handling unit yielded results of 0.78, 0.20, and 0.10 s/cm^3. Prior to this air sampling, furniture had been removed from the rooms and the floors had been cleaned using asbestos protocols.

Principal assumptions for this investigation were:

1. The debris found inside the duct was introduced several years previously, during the replacement of a reheat coil.

2. Asbestos fibers were uniformly distributed to all spaces served by the air-handling unit, and these fibers were released from the debris into the supply duct airstream in a steady-state fashion.

3. The asbestos-containing settled dust conditions in the three rooms where aggressive air sampling was conducted (test rooms) were representative of asbestos-containing settled dust conditions throughout the spaces served by the air-handling unit.

4. The three aggressive air samples were representative of all the spaces served by the air-handling unit.

5. All fibers present in the test rooms were suspended by the aggressive air sampling protocol and were available for capture by an air sampling cassette.

6. All fibers present in the test rooms during sampling came from the debris inside the duct.

7. The air handling unit had operated 24 hours per day, without significant interruption, from the time the reheat coil was replaced until the time that the three aggressive air samples were taken.

It was possible to estimate the average airborne asbestos concentration using the facts and assumptions stated above plus certain other data, such as the quantity of airflow from the air-handling unit to the test rooms. As a worst case scenario, the highest airborne concentration from the aggressive air sampling, 0.78 s/cm^3, was used as a base for this estimate. From this airborne concentration, the total number of asbestos structures present in the test room was estimated. Using the assumptions stated above and the measured areas of the various surfaces in the room, an asbestos loading per unit area was estimated for each surface. Knowing the total number of structures present and given the assumptions, the average airborne concentration of asbestos in the spaces served by the air-handling unit was estimated to be 1.9×10^{-6} s/cm^3. This calculated concentration was based on TEM analysis using the AHERA regulation protocol, which is more stringent than the analytical protocol specified by OSHA for worker exposure. (This estimate included a correction factor to compensate for the absence of furniture during the air sampling in the test room and for the fact that the floor had been cleaned in the test room prior to sampling.)

If assumption #5 was altered to postulate that only 1% of the fibers present in the test room was suspended by the aggressive air sampling and if it were further postulated that only 10% of the fibers which were released from the debris was still present in the test room at the time of aggressive air sampling, the estimate became 1.9×10^{-3} s/cm^3, which rounded is .002 s/cm^3. This level is an order of magnitude lower than the clearance level mandated by the AHERA regulation for reoccupancy of a school after asbestos removal (or other abatement activity) has occurred. It was the conclusion of the industrial hygienist, therefore, that exposure to occupants of the space served by the air-handling unit due to this debris incident was on the average very low.

Case Study #2

An industrial hygiene consultant conducted an evaluation of the significance of asbestos structures found in settled surface dust in a multi-story office building located in an urban area. The building was approximately twenty years old, with no major additions. The construction was primarily reinforced concrete floor slabs, columns, and concrete beams with a curtain wall exterior. Each floor occupied by the client was similar in layout and use of finish materials. The per floor area was approximately 24,000 square feet (SF). The facility was part of a larger complex of buildings. The project focused on specific floors occupied by the client.

The purpose of this investigation was to recommend procedures which would reduce the potential for disturbance of settled asbestos surface dust in the facility. This study was designed to investigate the type, distribution and quantities of asbestos in settled surface dust in floors of the facility occupied by the client; to recommend and design cleaning methods and work practices sufficient to clean the surfaces; in accordance with management's decision to clean designated floors, to recommend quality assurance practices to monitor the effectiveness of the cleaning procedures.

After a visual inspection of the facility and meetings with various parties, the I.H. consultant recommended an analytical protocol employing TEM to identify and quantify asbestos structures in settled surface dust in the facility, as well as in other buildings in geographic proximity (1/2 mile radius) to the facility, and at outdoor locations. Since the client had previously made a decision to conduct a special cleaning of the building in response to the presence of asbestos in settled dust, the I.H. consultant recommended three models for cleaning, representing increasing complexity and stringency, of settled surface dust. These cleaning procedures were designated as levels 1, 2, and 3, with level 3 being the most stringent. A validation testing procedure for each protocol was identified to determine the success of the cleaning at the facility. Along with each protocol, quality control procedures were specified. An investigative protocol was also established to quantify the levels of asbestos structures found in the facility as well as other similar and non-similar buildings located in the geographical area and the ambient environment.

The investigation concluded that the asbestos structure content of settled surface dust sampled in the facility was comparable to that found in buildings, from the geographic area, which did not have ACM's. Nearly all of the asbestos structures found in the facility were of the small chrysotile type, which are often ubiquitous in urban environments. Results of concurrently conducted ambient air monitoring data, including a review of the historical air monitoring data, showed that airborne asbestos structures were below the most stringent levels set by EPA (for schools).

The client decided to do at least some special cleaning, even though the settled dust asbestos levels were low, as a precaution and to actively demonstrate concern for the building occupants. Consequently, a recommendation to employ the level 1 cleaning method was made by the I.H. consultant. Those special cleaning procedures were tested in a controlled manner in an isolated area within the facility. Although the special cleaning was not classified as asbestos abatement work, the work was performed by a licensed asbestos abatement contractor. This was done in order to ensure the best possible results using trained, certified workers experienced with HEPA filtered vacuum cleaners and other special cleaning means and methods. All special cleaning work was performed over weekend/holiday periods. Air clearance testing by PCM was performed after the cleaning of each floor and before reoccupancy. Quality control air testing, using TEM, was used to routinely revalidate the efficiency of the level 1 cleaning procedures as they were performed and to make adjustments to the procedures, as necessary.

The conclusion of the I.H. consultant was that the total quantity of settled surface dust found within office spaces and contiguous areas of the facility had been reduced significantly from precleaning levels.

REFERENCES

1. **U.S. Environmental Protection Agency,** *Guidance for Controlling Asbestos-Containing Materials in Buildings,* EPA Office of Pesticides and Toxic Substances, Washington, DC, 560/5-85-024, June, 1985.
2. **National Institute of Building Sciences,** *Guidance Manual: Asbestos Operations and Maintenance Work Practices,* NIBS Document Number 5076-7, National Institute of Building Sciences, Washington, DC, 1992.
3. **Ganick, N.R. and Hays, S.M.,** HVAC modifications for asbestos operations and maintenance purposes, presented at Health Effects Institute, Cambridge, Massachusetts, March 8-9, 1993.
4. **Hays, S.M. and Millette, J.R.,** Use of observational data and experimental studies in developing better operations and maintenance plans, presented at Environmental Management '92, The Ninth Conference and Exposition of NAC/The Environmental Information Association, Pittsburgh, April 8, 1992. (Included in appendices.)

Glossary

abatement - actions taken to lessen or reduce the risk associated with asbestos-containing materials in buildings.

acetone fusing - collapsing of an MCE filter surface with acetone vapors.

acicular - the shape shown by an extremely slender crystal with cross-sectional dimensions which are small relative to its length, i.e., needle-like.

adhesive tape lift - a method of collecting particles from surfaces for subsequent examination by microscopy. This sampling technique involves the use of various sticky films to which particles adhere.

aerosol - a dispersion of solid particles or liquid droplets in a gaseous medium.

ambient - refers to the surrounding area or environment.

ambient air - the surrounding air or atmosphere in a given area under normal conditions.

amphibole - a group of rock-forming ferromagnesium silicate minerals, closely related in crystal form and composition, and having the nominal formula:

$A_{0\text{-}1}B_2C_5T_8O_{22}(OH,F,Cl)_2$, where:
A = K, Na;
B = Fe^{2+}, Mn, Mg, Ca, Na;
C = Al, Cr, Ti, Fe^{3+}, Mg, Fe^{2+};
T = Si, Al, Cr, Fe^{3+}, Ti.

In some varieties of amphibole, these elements can be partially substituted by Li, Pb, or Zn. An amphibole is characterized by a cross-linked double chain of Si-0 tetrahedra with a silicon:oxygen ratio of 4:11, by columnar or fibrous prismatic crystals and by good prismatic cleavage in two directions parallel to the crystal faces and intersecting at angles of about 56° and 124°.

amphibole asbestos - amphibole in an asbestiform habit.

analytical sensitivity - the calculated settled dust asbestos structure concentration in structures/area, equivalent to the counting of one asbestos structure in the analysis.

architectural finish - a material not essential to the building structure, but which is applied to a building surface for the purpose of ornamentation and occupant comfort, e.g. plaster, wallcovering, paint, carpet.

asbestiform - a specific type of mineral fibrosity in which the fibers and fibrils possess high tensile strength and flexibility.

asbestos - a term applied to a group of silicate minerals belonging to the serpentine and amphibole groups which have crystallized in the asbestiform habit, causing them to be easily separated into long, thin, flexible, strong fibers when crushed or processed. The Chemical Abstracts Service Registry Numbers of the most common asbestos varieties are: chrysotile (12001-29-5), crocidolite (12001-28-4), grunerite asbestos (amosite) (12172-73-5), anthophyllite asbestos (77536-67-5), tremolite asbestos (77536-68-6) and actinolite asbestos (77536-66-4).

asbestos-containing material (ACM) - any material or product which contains more than 1 percent asbestos.

Asbestos Hazard Emergency Response Act (AHERA) - an Act of the United States Congress, signed into law on October 22, 1986, which resulted in an EPA regulation published in the October 30, 1987, *Federal Register* covering asbestos-containing materials in schools.

asbestos management plan - a strategy for managing asbestos in buildings by using a variety of techniques to prevent or properly control the release of asbestos fibers into the air until the ACM's are removed or the building is properly demolished. The AHERA regulation specifies management plan requirements for schools.

asbestos structure - a term applied to any connected or overlapping grouping of asbestos fibers or bundles, with or without other particles.

aspect ratio - the ratio of length to width of a particle.

background - the totality of effects that tend to obscure a phenomenum under investigation and above which measurements of the phenomenon must occur.

baseline - a value that represents the background level of a measurable quantity and is used as a basis for subsequent comparison or control.

blank - an unused sample collection item, such as a filter or wipe, analyzed to determine if contamination exists on the sample collection medium prior to collection of samples.

building inspection - the process of determining whether or not a building is constructed or equipped with materials that are suspected to contain potentially hazardous material.

bulk sampling - the technique used to collect samples, representative of the in-place materials, from suspect materials such as fireproofing, pipe and boiler insulation, and acoustical spray.

chrysotile - a fibrous mineral of the serpentine group which has the nominal composition:

$$Mg_3Si_2O_5(OH)_4$$

Most natural chrysotile deviates little from this nominal composition. In some varieties of chrysotile, minor substitution of silicon by Al^{3+} may occur. Minor substitution of magnesium by Al^{3+}, Fe^{2+}, Fe^{3+}, Ni^{2+}, Mn^{2+}, and Co^{2+} may also be present. Chrysotile is the most prevalent type of asbestos used in the United States.

clearance - the level below which air is determined to be "clean." For asbestos, clearance levels established by AHERA are 70 s/mm^2 by TEM (.01 f/cm^3 by PCM for limited situations).

cleavage - the breaking of a mineral along one of its crystallographic directions.

cleavage fragment - a fragment of a crystal that is bounded by cleavage faces.

cluster - a structure in which two or more fibers, or fiber bundles, are randomly oriented in a connected grouping.

coefficient of variation (CV) - a relative measure of dispersion found by expressing the standard deviation as a percentage of the mean.

computer-controlled scanning electron microscopy (CCSEM) - a scanning electron microscope interfaced with a computer which is programmed to move automatically from particle to particle to collect data about the particle and store in a computer file.

concentration - the amount of a particular agent or contaminant in a given volume of air.

conductive tape - tape that is made of a material that conducts electricity.

cyclone system - an inertial separator that imparts a centrifugal force to the particles to be removed.

decontamination - removing or inactivating polluting or harmful substances.

debris - particles of building materials which by color, texture, and appearance can be determined to have come from a particular source.

delamination - separation of one layer from another.

dimethyl formamide (DMF) - organic solvent used to dissolve MCE filters.

dust vacuum method (DVM) - a method of sampling interior surface dust by collecting the dust with a personal monitoring pump connected by Nalgene™ tubing to a three-piece air-monitoring cassette with a 0.8 μm MCE filter.

electron diffraction (ED) - a technique in electron microscopy by which the crystal structure of a specimen is examined.

electron microscope - a type of microscope that uses an electron beam and magnetic lens to produce enlarged images of objects.

encapsulation - the coating of asbestos-containing material with a bonding or sealing agent to prevent the release of fibers to the air.

enclosure - a resilient structure, built or sprayed around a particular agent, designed to prevent disturbance and contain a release of that agent.

energy dispersive X-ray analysis (EDXA) - measurement of the energies and intensities of X-rays by use of a solid state detector and multi-channel analyzer system for the purpose of elemental identification.

fiber - an elongated particle which has parallel or stepped sides. A fiber is defined by the AHERA regulation to have an aspect ratio equal to or greater than 5:1 and a minimum length of 0.5 μm.

fiber bundle - a structure composed of parallel, smaller diameter fibers attached along their lengths. A fiber bundle may exhibit diverging fibers at one or both ends.

fiber release - the generation of airborne fibers from an asbestos-containing source.

fibril - a single fiber of asbestos, which cannot be further separated longitudinally into smaller components without losing its fibrous properties or appearances.

fibrous structure - a fiber, or connected grouping of fibers, with or without other particles.

field blank - a sample container which has been taken to the sampling site, opened, and then closed. Such a container is used to determine the background structure count for the measurement.

forensic evidence tin - metal dish with tight-fitting lid.

HEPA vacuum - a high-efficiency particulate air (HEPA) filtered vacuum capable of trapping and retaining at least 99.97 percent of 0.3 micrometer, and larger, diameter mono-dispersed particles.

hexane - (C_6H_{14}), a colorless, volatile alkane that is toxic and flammable, used as a solvent and paint diluent.

in-place material - building products which exist as installed.

Jaffe washer - petri dish with organic solvent used to dissolve filter material in the preparation of grids for TEM analysis.

limit of detection - the calculated asbestos structure concentration equivalent to counting of 2.99 asbestos structures in the analysis.

management in-place - a formulated plan of training, special work practices, and surveillance intended to maintain ACM's in good condition. This concept is realized in practice by use of an operations and maintenance plan.

matrix - a structure in which one or more fibers, or fiber bundles, touch, are attached to, or are partially concealed by, a single particle or connected group of non-fibrous particles.

methyl cellulose - a surfactant used to disperse particles.

microscopy - science involved with magnifying objects for examination.

microvac sampling - a method of collecting particles from surfaces for subsequent examination by microscopy. This sampling technique involves the use of filter cassette and an air sampling pump.

mixed cellulose ester (MCE) filter - cellulose strands bound together in a web called "tortuous pore", which display a very irregular surface when observed under magnification.

operations and maintenance (O&M) - specific procedures and practices developed for the interim control of asbestos-containing materials in buildings until they are removed. An O&M plan is part of a comprehensive asbestos management plan.

particulate - a material composed of separate and distinct particles.

particle dispersion technique - a procedure for moving apart particles which are close together or piled on top of each other, for subsequent analysis.

passive dust canister - container placed in such a way to collect settling dust over a specific period of time.

passive monitor - a collection device which is placed in such a way to accumulate a sample without the need for an external power supply.

phase contrast microscopy (PCM) - an optical microscopic technique used for the counting of fibers in air samples, but which does not distinguish fiber types.

PCM equivalent fiber - a fiber of aspect ratio greater than or equal to 3:1, longer than 5 µm, and which has a diameter between 0.2 µm and 3.0 µm.

PCM equivalent structure - a fibrous structure of aspect ratio greater than or equal to 3:1, longer than 5 μm, and which has a diameter between 0.2 μm and 3.0 μm.

plasma asher - instrument which uses an oxygen plasma to ash materials at low temperature.

plasma etching - process of ashing the top layer of a filter in a plasma asher.

plenum - a horizontal space designed to transport air in a building; commonly the space between a suspended ceiling and the floor above.

polarized light microscopy (PLM) - an optical microscopic technique used to distinguish between different types of asbestos fibers by their shape and unique optical properties.

polychlorinated biphenyls (PCBs) - any of a group of toxic, chlorinated aromatic hydrocarbons used in a variety of commercial applicants such as paints, inks, adhesives, electrical condensers, batteries, and lubricants.

Post-it™ note sampling - a method of collecting particles from surfaces for subsequent examination by microscopy. This technique involves the use of commercially available Post-it™ notes.

primary structure - a fibrous structure that is a separate entity in the TEM image.

quantification - the process of expressing something as a number or amount.

removal - the taking out or stripping of asbestos, tremolite, anthophyllite, or actinolite or materials containing asbestos, tremolite, anthophyllite, or actinolite.

replacement - installing non-ACM's after removal of ACM's.

replication - a procedure in electron microscopy specimen preparation in which a thin copy, or replica, of a surface is made.

risk - the potential to release asbestos fibers into the air or to otherwise cause fibers to become airborne.

sampling - collecting particles from surfaces for subsequent examination by microscopy.

scanning electron microscopy (SEM) - a method of microscopic analysis which directs an electron beam at the sample and then collects the beams that are reflected to produce an image from which fibers can be identified and counted.

scrape and scoop - a method of collecting particles from surfaces for subsequent examination by microscopy. This sampling technique involves scraping together a pile of dust by using a card and scooping it into a container that can be closed tightly.

selected area electron diffraction (SAED) - a technique in electron microscopy in which the crystal structure of a small area of a sample is examined.

serpentine - a group of common rock-forming minerals having the nominal formula:

$$Mg_3Si_2O_5(OH)_4$$

settled dust - particulate material less than one (1) mm in its longest dimension, which is deposited on a surface.

surface contamination - the fouling of a surface with an undesired or unexpected chemistry; in the assessment of buildings, the part of the dust which contains materials that are considered toxic.

surface loading - the quantification of asbestos structures which are present per unit area of a given surface.

surfactant - a chemical wetting agent added to water to improve its penetration abilities into asbestos-containing materials.

suspect materials - materials suspected, but not known, to contain asbestos. Actual determination must be made by instrumental analysis.

suspension - a system in which very small particles of solid, semisolid, or liquid material are more or less evenly dispersed in a liquid or gas phase.

tape sampling - a method of collecting particles from surfaces for subsequent examination by microscopy. This technique involves the use of viscid films like cellophane tape.

tared - pre-weighed.

thermal systems insulation (TSI) - ACM applied to pipes, boilers, tanks, ducts, or other building components to prevent heat loss or gain or water condensation.

toxic - relating to or caused by a poison or toxin.

transmission electron microscopy (TEM) - a method of microscopic analysis which utilizes an electron beam that is focused onto a thin sample. As the beam penetrates (transmits) through the sample, the difference in densities produces an image on a fluorescent screen from which samples can be identified and counted.

transuranics - elements with an atomic number greater than 92, the atomic number of uranium.

ultrasonic - having to do with very high frequencies of acoustic energy, above the upper range of human hearing (approximately 20,000 hertz).

unopened fiber - a large diameter asbestos fiber bundle which has not been separated into its constituent fibrils or fibers.

vermiculite - an expanded mica (hydrated magnesium-aluminum-iron silicate) used in lightweight aggregates, insulation, fertilizer, and soil conditioners, as a filler in rubber and paints, and as a catalyst carrier.

visual assessment - observation via the naked eye, i.e. no magnification.

wet-wipe sampling - a method of collecting particles from surfaces for subsequent examination by microscopy. This technique involves the use of cloths moistened with filtered water or commercially pre-moistened cloths such as baby wipes.

ABBREVIATIONS

ASHAE - American Society of Heating and Air-Conditioning Engineers, Inc. (now ASHRAE)
ASHRAE - American Society of Heating, Refrigerating and Air-Conditioning Engineers, Inc.
ASTM - American Society for Testing and Materials
EPA - U.S. Environmental Protection Agency
HUD - U.S. Department of Housing and Urban Development
NIBS - National Institute of Building Sciences
NIEHS - National Institute of Environmental Health Sciences
NIOSH - National Institute for Occupational Safety and Health
OSHA - Occupational Safety and Health Administration

SYMBOLS

dpm - disintegrations per minute
µg - microgram (10^{-6} grams)
µm - micrometer (10^{-6} meters)
f/cm^3 - fibers per cubic centimeter
f/cm^2 - fibers per square centimeter
l/min - liters per minute
s/cm^3 - structures per cubic centimeter
s/cm^2 - structures per square centimeter

APPENDICES

APPENDIX 1

Asbestos Exposure During and Following Cable Installation in the Vicinity of Fireproofing

William M. Ewing,[A] Jean Chesson,[B] Tod A. Dawson,[A] Eva M. Ewing,[A] Richard L. Hatfield,[C] Steve M. Hays,[D] Dale L. Keyes,[E] William E. Longo,[F] James R. Millette,[G] and William H. Spain[A]

A study was conducted to determine whether installing cables above a suspended tile ceiling where a low-density asbestos-containing fireproofing was present on the structural beams and decking would result in elevated airborne concentrations of asbestos. The study was designed to simulate routine cable installation and cleaning typical of that used by service workers when they are unaware of the presence of asbestos. Personal and area air sampling was conducted before, during and after the installation of 12 cables. Personal and area air sampling was also conducted during cleaning activities approximately two hours after the cable installation. All samples and blanks were analyzed by transmission electron microscopy (TEM) using an indirect preparation technique. Results are reported as asbestos structures per cubic centimeter of air (s/cm^3).

The geometric mean of area samples collected prior to cable installation was 0.002 s/cm^3. The mean of area samples collected during the cable installation was 3.5 s/cm^3 and the mean of personal samples was 26 s/cm^3. Area sampling conducted during a period following completion of the cable installation averaged 3.4 s/cm^3. Airborne asbestos during the clean-up phase of the study averaged 55 s/cm^3 (area samples) and 44 s/cm^3 (personal samples). All of the increases in asbestos concentrations were statistically significant when compared to levels measured before the work began.

The results demonstrated a significant increase in airborne asbestos concentrations during and immediately following cable installation activities compared to concentrations before cable installation. The results also demonstrate a significant increase in airborne asbestos concentrations during routine cleaning activities.

INTRODUCTION AND BACKGROUND

The presence of friable asbestos-containing materials (ACM) in buildings is of interest due to the potential for exposure of building occupants and workers to asbestos fibers.[1] Asbestos exposures fall into two categories: prevalent level exposures and episodic exposures.[2] Episodic exposures frequently result from custodial or maintenance activities which directly disturb in-place ACM or dust and debris derived from these materials.[1–6] For service workers who routinely work in the vicinity of ACM, episodic events may characterize their prevalent exposure.

Service workers who perform custodial and maintenance tasks in buildings with friable ACM have been shown to be at greater risk of exposure and disease.[5,7–12] Only a few studies, however, have been conducted to determine what level of exposures occur during tasks that service workers perform.[11–16] Accordingly, a series of experiments were designed and conducted to gather additional data to help characterize these exposures. This article reports on one of these experiments.

One common activity which occurs in office buildings is the installation of cables and wiring for communications and electrical service. In buildings with suspended ceilings, cables and wires are often installed above the suspended ceiling systems. Exposure to airborne asbestos may occur when asbestos-containing fireproofing is present above the suspended ceiling and when dust and/or debris from the fireproofing is present on the ceiling tiles.

A related study previously investigated whether installing cable in a ceiling plenum where asbestos-containing fireproofing was present results in elevated airborne asbestos concentrations.[14] The results of that study indicated a 500-fold increase during the work activities with several samples exceeding 50 asbestos structures per cubic centimeter (s/cm^3) as measured with transmission electron microscopy (TEM). This previous study was performed in a school building with fireproofing containing approximately 10–12% chrysotile asbestos (by weight) in a gypsum and vermiculite matrix.

The purpose of the present study was to determine whether similar results are obtained with a different type of asbestos-containing fireproofing. The previous study involved a wet-applied high-density material. The current study involved a lower density material composed of approximately 25–30% (by weight) chrysotile asbestos in a mineral wool (60%), cement (2%), and bentonite (3%) matrix. The study was designed to evaluate the methods historically used by service workers installing cable or used today by individuals either not aware that asbestos-containing materials are present or unwilling to take proper precautions. Despite the attention given to asbestos in recent years, we have found that the failure to take precautions when working around ACM is not uncommon.

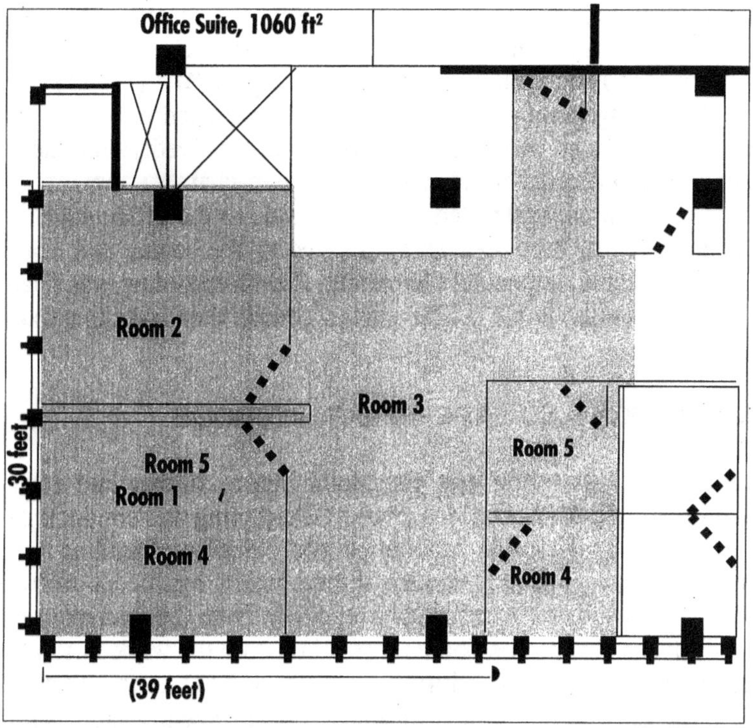

Figure 1. Diagram depicting test site.

TEST SITE

The test site consisted of a five-room vacant office suite encompassing 1,060 square feet (ft^2) (Figure 1). The site was located on the sixth floor of a 30-story building constructed in 1965–1966. No furnishings were present in the office suite. The suite consisted of a central room and four adjoining smaller offices. The ceiling system was composed of 2′ × 4′ lay-in mineral fiber tiles supported by a suspended T-bar grid. No asbestos was detected in three bulk samples of the ceiling tiles. The walls between each office were partitions which did not extend above the suspended ceiling. The distance from the suspended ceiling to the overhead deck was approximately

26 inches. The distance from the suspended ceiling to the bottom of the flanges of beams was approximately 11.5 inches and from the suspended tile ceiling to the floor was approximately 8.5 feet.

In preparation for the study, openings between the office suite and adjoining areas were sealed with 6-mil polyethylene sheeting. These included all heating, ventilation and air conditioning (HVAC) supply air diffusers in the ceiling and one wall mounted return air gill. The HVAC system did not use the ceiling space as an air plenum. The carpeted floor and two grass cloth wall coverings were covered with 6-mil polyethylene sheeting to prevent contamination of these materials. A two-chamber personnel decontamination unit was constructed at one entrance to the office suite.

The fireproofing had been spray applied to the beams and corrugated steel deck above the suspended tile ceiling. The fireproofing was friable and dust and debris was visible on the tops of ceiling tiles and light fixtures.

STUDY DESIGN AND EXPERIMENT METHODS

Stationary air sampling was conducted before, during and after cable installation and cleaning activities. Personal sampling was conducted on two individuals during the cable installation and cleaning phases of the study. The study was designed to determine if there was a significant difference in airborne asbestos fiber concentrations resulting from the different activities or phases of the study. The phases were (1) before the work simulation, (2) during cable installation, (3) during cleaning, and (4) during a "die-down" period between installation and cleaning. A sufficiently large number of samples were collected so there would be a high probability of observing a statistically significant increase if the actual increase were 10-fold or greater.

CABLE INSTALLATION PROCEDURES

Twelve cables were cut to the desired length, labeled, coiled and placed in room 4 prior to beginning the study. Twelve cables would be sufficient to install twelve work stations. Number 16/3 electrical cord was used as the cable. The method of installing the cables was consistent with the method used in the electrical trade and with procedures used previously in the building. This method is referred to as the "throw and pull" method. A weight is attached to a string, which is attached to the cable, and is then thrown in the desired direction. The weight used was a one-inch steel nut.

One worker in room 4 threw the nut through the ceiling space to an open tile which was previously removed by a second worker. The second worker then pulled the string and attached cable to the opening. A typical throw was six to eight feet. The process was repeated as necessary for longer cable runs. The two individuals were instructed to avoid any direct contact with the fireproofing material and to avoid damaging the ceiling tiles or other building components to the extent possible. Figures 2 and 3 depict the cable installation process.

The installation of 12 cables required removal of 14 of the 134 ceiling tiles (10.4%) present in the office suite. Each tile was carefully removed from the ceiling grid, lowered to the floor, and stacked along the wall of each room. The location of each cable installation is shown on Figure 4.

The cable installation required 57 minutes to complete. The cables were then removed and the ceiling tiles replaced (26 minutes). Air sampling was only conducted during the installation phase since this was the specific activity of interest. A die-down period of 109 minutes followed with the resumption of area air sampling to serve as a baseline for comparison with the samples collected during the cleaning phase. Immediately following the die-down period, cleaning of the office suite was performed during a 30-minute period with area and personal air sampling conducted during this phase.

The routine cleaning consisted of two individuals sweeping the floor of each room with brooms, collecting any visible dust or debris in a dustpan, and depositing it into a labeled asbestos disposal bag. The few horizontal surfaces (e.g., door ledges) were wiped with a dry cotton cloth. Cleaning activities are depicted in Figure 5.

All participants in the study, including cable installers, clean-up personnel and observers wore air-purifying respirators and full-body protective clothing during all phases of the study. A two-chamber decontamination station employing a high efficiency particulate air (HEPA) filtered vacuum and wet cleaning was used for personnel and equipment exiting the work area. The HEPA filtered vacuum remained operating during the study to provide a slightly reduced pressure inside the work area with respect to surrounding areas.

SAMPLING AND ANALYTICAL METHODS

Area air sampling was conducted at stationary locations depicted on Figure 4. Additional area air sampling was conducted outside the office suite as a check for fiber migration. Personal sampling was conducted in the breathing zone of two individuals during the cable installation and cleaning activities. Area samples were collected using high-volume air sampling

Figure 2. View of cable installation process from floor level.

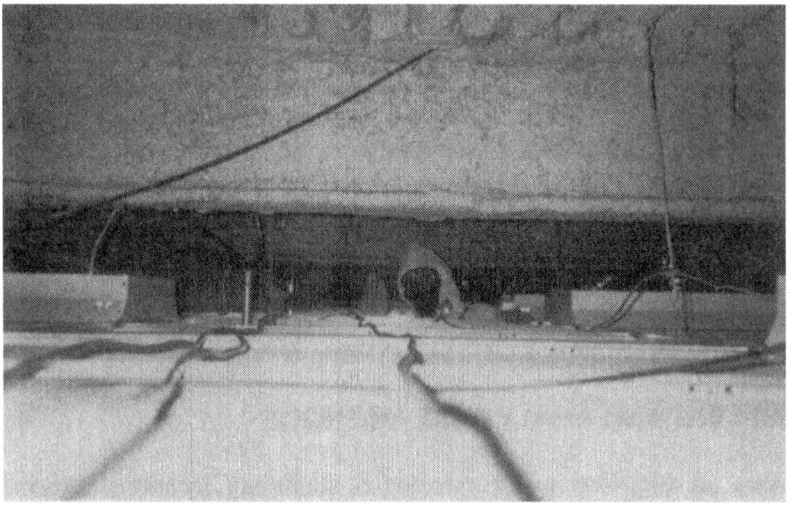

Figure 3. View of installed cables laying atop ceiling tiles with spray-applied fireproofing on the beam and deck above.

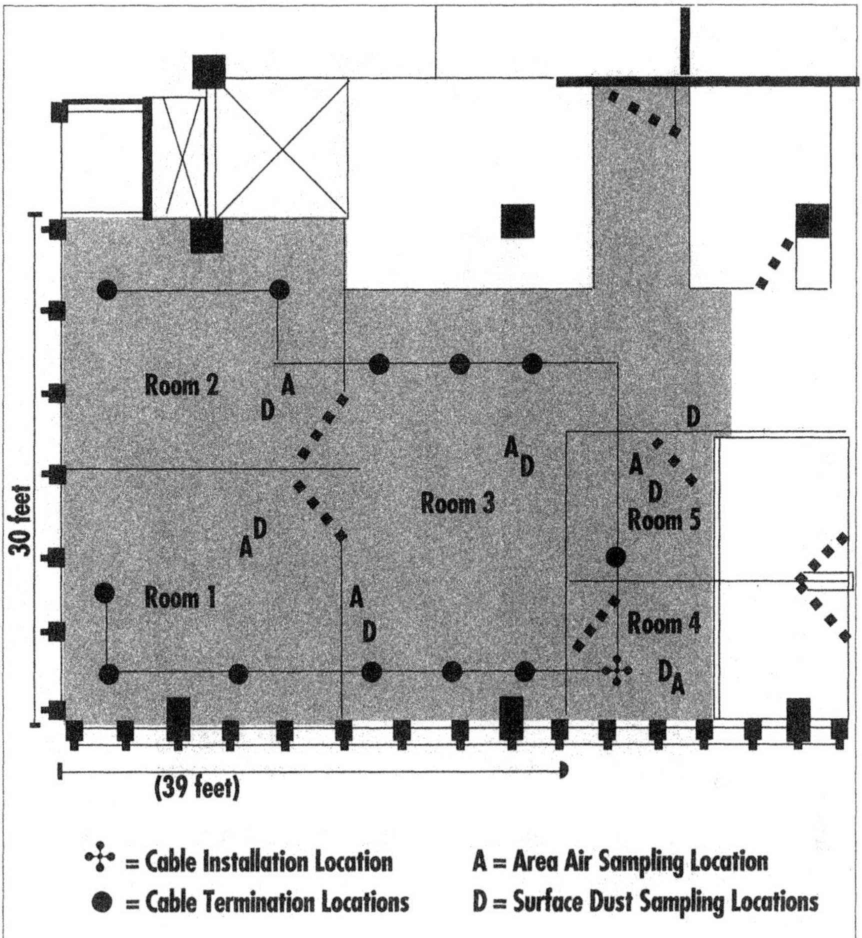

Figure 4. Diagram depicting cable routing and sampling locations.

pumps at nominal flow rates of 10 liters per minute (8.9–10.1 lpm). Personal samples were collected using personal sampling pumps at nominal flow rates of 2 lpm (1.9–2.0 lpm). All air sampling pumps were calibrated before and after sampling using a Mini-Buck, model M-30 automated soap bubble calibrator. The sampling media consisted of a 0.45 micrometer (μm)

Figure 5. View of clean-up procedures with area air sampling pump at right.

pore size mixed cellulose ester (MCE) filter in a 25 mm sampling cassette as described in 40 CFR 763, Subpart E, Appendix A.[17]

Surface dust samples were collected from the floor (plastic covering on carpet) as locations shown on Figure 4. The locations coincided with the area air sampling locations. Surface sampling was performed following the cable installation and die-down period but before cleaning activities began. Surface sampling was performed using the microvacuum technique.[18] The surface collection area was 100 cm^2 for each sample. The sample was collected using a 0.45 µm MCE filter in a closed-face 25 mm diameter sampling cassette with a two-inch cowl. A 2.5 cm length of Tygon™ tubing cut with a serrated end was attached to the cassette inlet. Flow rate through the cassette was maintained using a personal sampling pump calibrated at

2.0 lpm. The surface dust of each sampled area was collected by passing the nozzle across the sampling area twice for a total period of one minute.

At the conclusion of each air or dust sample, the sampling cassette was capped and sealed with tape. Four field blanks were collected and analyzed concurrent with the field samples. All samples were stored and hand-carried upright to the analytical laboratories. Six area air samples were typically collected during each phase of the study. Five from the six were randomly selected for TEM analysis. The remaining sample from each set was stored in case a sample was voided due to damage. No samples were voided. All personal samples and all surface dust samples collected after the cable installation were submitted for analysis. Each of the samples were analyzed using TEM with indirect preparation.

The indirect preparation technique is preferred when comparing results from similar data sets.[19–20] The method has been previously described by Yamate, et. al.[21] The relative merits of direct and indirect preparation techniques are still being debated. Nevertheless, it is apparent that the indirect preparation technique is particularly suited for experimental studies in which measured concentrations may vary over a wide range. Identification of asbestos structures was based on morphology, selected area electron diffraction (SAED), and energy dispersive X-ray analysis (EDXA). All asbestos structures that could be identified were counted.

Laboratory and field blanks were prepared and analyzed along with the samples. Twenty percent of the air samples analyzed by the primary laboratory were randomly selected for quality control (QC) analysis by a second laboratory.

DATA ANALYSIS

The geometric mean was used to represent the average or typical concentration of airborne asbestos fibers for each set of samples. Air pollution data have been shown to be lognormally distributed.[22] Accordingly, log transformation of the data results in a good approximation of a normal distribution and allows for the use of standard statistical tests. The mean of the log transformed data (geometric mean) is the appropriate measure of central tendency.

Analysis of variance was used to examine differences in the geometric means among the various phases of the study, with the Scheffé test used to determine the statistical significance of the observed differences.[23] The results are expressed as *p* values, with small values indicating low probabilities that the observed differences could have occurred by chance alone. A *p* value smaller than 0.05 is usually considered "statistically significant."

Table 1. Simulation of cable installation. Summary of air sampling results.

Phase	Arithmetic Means (s/cm^3)	Arithmetic Standard Deviation (s/cm^3)	Geometric Mean (s/cm^3)	Number of Observations
Before Installation	0.006	0.014	0.002	5
During Installation (Area Samples)	3.6	0.84	3.5	5
During Installation (Personal Samples)	26	7.5	26	2
Die-Down	3.8	1.9	3.4	5
Cleaning (Area Samples)	58	19	55	5
Cleaning (Personal Samples)	45	7.9	44	2
Background (Outside Worksite)	0.016	0.028	0.004	3

PRESENTATION OF RESULTS

The results of all area and personal samples analyzed are summarized in Table 1. The geometric mean asbestos concentration (area samples) before the study was 0.002 s/cm^3. During the cable installation period the mean concentration increased to 3.47 s/cm^3. The mean concentration for two personal samples collected on the installers was 26 s/cm^3. The geometric mean value for area samples collected during the die-down period following cable installation was 3.4 s/cm^3.

During cleaning of the office suite after the die-down period the geometric mean increased to 55 s/cm^3 as reflected by area samples. Personal samples collected during the same 30-minute cleaning period on the two individuals averaged 44 s/cm^3. The set of samples collected outside the office suite during cable installation and cleaning as a check for contamination indicated a geometric mean of 0.004 s/cm^3.

The results of the comparison of geometric means are presented in Table 2. These comparisons indicate a clear statistically significant increase during cable installation when compared to samples collected before the study began (<0.00005). In addition, a significant increase was found when cleaning activities were conducted after the die-down period (p=0.0023).

The arithmetic and geometric mean values for five settled dust samples taken from the floor surface following cable installation was 900,000 s/cm^2 and 190,000 s/cm^2, respectively. The difference between these means indicates a non-homogeneous deposition of dust across the floor surface. The settled dust represents an accumulation during the cable installation and die-down period phases of the study (192 minutes).

Analysis of the asbestos structures identified in 19 area and personal air samples indicated a composition of 58% free fibers, 21% fiber bundles, 15% matrices, and 6% clusters. For all structures counted, 8% were greater than 5 micrometers in length.

The results of quality assurance comparisons for six samples analyzed by two laboratories indicated a Pearson correlation coefficient of 0.97. The quality control samples were selected in a stratified random fashion to capture a broad range of concentrations. In addition to being highly correlated, analyses of the six samples revealed good agreement in absolute terms and no pattern of higher values for either one of the laboratories. Most importantly, the conclusions of the study would have been the same regardless of which laboratory analyzed the samples.

DISCUSSION

The results of this study involving a low-density fireproofing are similar to those obtained previously with a high-density fireproofing.[14] In each case, elevated airborne asbestos concentrations resulted from installing cable and the associated cleaning activities.

Accordingly, it may be concluded that cable installation involving traditional methods in the vicinity of asbestos-containing fireproofing can cause elevated asbestos exposures for those performing the task and others in the vicinity of the work.

This study was not designed to compare exposure levels to current or previous OSHA permissible exposure limits (PELs). Furthermore, TEM and phase contrast microscopy (PCM) measurements are not directly comparable. However, some feeling for the likely magnitude of a concentration if

Table 2. Simulation of cable installation. Comparison of geometric means (Scheffé Test).

Comparison[1]	Significance (p value)
Before vs. Installation	<0.00005
Before vs. Cleaning	<0.00005
Before vs. Die-Down	<0.00005
Before vs. Background	0.9630
Cleaning vs. Installation	0.0024
Cleaning vs. Die-Down	0.0023

[1] All other comparisons were not significant at the p=0.05 level.

it were measured by PCM can be obtained by considering only structures longer than 5 μm and wider then 0.25 μm. Using this approach, the personal samples collected in this study would have a PCM equivalent value in the range 0.5–2 f/cm^3 if no interfering fibers were present. However, since the fireproofing involved in this study contained 60% mineral wool, PCM values would be a poor surrogate for actual asbestos exposure. The relationship between TEM and PCM measurements is discussed in detail in Chesson, Rench, Schultz and Milne (1990).[24]

Identifying the likely sources of exposure is important if exposures are to be controlled. During cable installation the primary source was asbestos-containing dust and debris on the top of 14 ceiling tiles removed to accommodate the maintenance activity. The secondary source was the in-place fireproofing itself. During the cleaning phase, the only significant source was the asbestos dust and debris on the floor and other horizontal surfaces within the office suite. Accordingly, control measures should focus on the asbestos-containing dust and prevention of its resuspension. For example, precleaning of ceiling tiles prior to removal should help reduce exposures. Further research is clearly needed in this area to evaluate alternative control techniques and work practices.

These data suggest that cable installers and custodians have in the past experienced episodic exposures such as these demonstrated. Several articles in the literature have demonstrated an increase in asbestos-related disease among building service workers, maintenance personnel, and custodians.[(7–9,25–26)] To date, however, little information is available regarding asbestos related disease among communication workers and electricians.[27] The need exists to characterize past exposure to these population groups and through epidemiological investigation determine to what extent these groups are at risk of developing an asbestos related disease.

Table 3. Area and personal sampling results.

Sample Number	Description	Concentration (s/cm^3)*
1002	Before Cable Installation, Room 1	<0.016
1003	Before Cable Installation, Room 2	<0.016
1004	Before Cable Installation, Room 3	<0.016
1005	Before Cable Installation, Room 5	0.032
1006	Before Cable Installation, Room 4	<0.016
1007	Cable Installation, Room 4	3.9
1008	Cable Installation, Room 1	3.6
1010	Cable Installation, Room 2	2.2
1011	Cable Installation, Room 3 (South)	4.5
1012	Cable Installation, Room 3 (North)	3.6
1017	Personal Sample, Feeding Cable	32
1018	Personal Sample, Receiving Cable	21
1021	Die-Down Period, Room 5	5.1
1022	Die-Down Period, Room 3 (South)	2.0
1023	Die-Down Period, Room 2	1.9
1024	Die-Down Period, Room 3 (North)	4.0
1025	Die-Down Period, Room 1	6.2
1026	Cleaning, Room 5	57
1027	Cleaning, Room 3 (North)	68
1028	Cleaning, Room 3 (South)	80
1029	Cleaning, Room 4	51
1031	Cleaning, Room 2	30
1033	Personal Sample, Cleaning	39
1034	Personal Sample, Cleaning	50
1014	Perimeter Sample, Outside Test Site During Cable Installation and Cleaning	<0.024
1015	Perimeter Sample, Outside Test Site During Cable Installation and Cleaning	<0.026
1016	Perimeter Sample, Outside Cable Test Site During Installation and Cleaning	0.049

* Results include all structures detected. Chrysotile was the only form of asbestos found.

Studies of this type further demonstrate the need to evaluate the sampling and analytical protocols to measure episodic exposures. For example, should phase contract microscopy be used to measure airborne asbestos with its inherent inability to distinguish between asbestos and nonasbestos fibers or to resolve fibers thinner than 0.25 μm and (based on current PCM counting procedures) shorter than 5 μm? To do so would eliminate approximately 90% or more of the asbestos fibers available for exposure from consideration and permit interfering fibers (cellulose, mineral wool, synthetic fibers) to further cloud the result.[28–30]

Dust sampling provides useful information for designing measures to reduce exposures. Air sampling can only document the presence of asbestos in the air. It cannot detect potential sources of airborne asbestos such as dust. Surface dust sampling permits sources of asbestos available for re-entrainment to be pinpointed and remedial measures taken before the actual exposure occurs. Further research is still necessary to standardize dust and air sampling techniques, including the number of samples necessary to characterize a given surface area, recovery efficiency from various surfaces, and the analytical technique employed when heavy concentrations of asbestos and other particles are encountered.

ACKNOWLEDGMENT

This study was funded, in part, by the Plaintiff's Executive Committee of the National School Asbestos Class Action. The authors retained full control over the design and execution of the study, and the interpretation and reporting of the results. The authors are grateful to Bonnie Bonham for her work in preparing the manuscript.

REFERENCES

1. **U.S. EPA Office of Pesticides and Toxic Substances:** *Guidance for Controlling Asbestos-Containing Materials in Buildings.* USEPA Pub. No. 560/5-85-024, Washington, DC (1985).
2. **Lory, E.E. and D.S. Coin:** Technical Report R883, "Management Procedure for Assessment of Friable Asbestos Insulating Material. Civil Engineering Laboratory," Naval Construction Battalion Center, Port Hueneme, CA (1981).
3. **Selikoff, I.J.:** The Third Wave of Asbestos Disease: Exposure to Asbestos in Place. Andrews Communications, Inc., New York (1990).
4. **Department of the Environment:** Asbestos Materials in Buildings, 2nd Edition. Her Majesty's Stationary Office, London, UK (1986).

5. *Managing Asbestos in Place: A Building Owner's Guide to Operations and Maintenance Programs for Asbestos-Containing Materials.* USEPA Pub No. 20T-2003, Washington, DC (1990).
6. **Millette, J.R., W.M. Ewing and R. Brown:** "Stepping on Asbestos Debris." *Microscope,* pp. 321–326 (1990).
7. **Oliver, L.C., N.L. Sprince, and R. Greene:** "Asbestos-Related Disease in Public School Custodians." *Am. J. Ind. Med.* 19:303–316 (1991).
8. **Huncharek, M., J. Muscat, and J. Capotorto:** Pleural Mesothelioma in a Lift Mechanic. *Br. J. Ind. Med.* 46:500–501 (1989).
9. **Stein, R.C., J.Y. Kitajewska, J.B. Kirkham, N. Tait, G. Sinha, and R.M. Rudd:** Pleural Mesothelioma Resulting from Exposure to Amosite Asbestos in a Building. *Resp. Med* 83:237–239 (1989).
10. **Litzistorf, G., M.P. Guillemin, P. Buffat, and F. Iselin:** Influence of Human Activity on the Airborne Fiber Level in Paraoccupational Environments. *J. Air Poll. Control Assoc.* 35(8):836–837 (1985).
11. **Paik, N.W., R.J. Walcott, and P.A. Brogan:** Worker Exposure to Asbestos During Removal of Sprayed Material and Renovation Activity in Buildings Containing Sprayed Material. *Am. Ind. Hyg. Assoc. J.* 44(6): 428–432 (1983).
12. **Sawyer, R.N.:** Asbestos Exposure in a Yale Building. *Environ. Res.* 13: 146–169 (1977).
13. **Crandlemere, R.W., D. Azrael, M.S. Evans, and S. Brown:** PCM and TEM Evaluation of Disturbance Testing of Asbestos "Popcorn" Type Acoustical Ceilings. *Nat. Asbestos Council J.* 7(2):47–51.
14. **Keyes, D.L., J. Chesson, and W.M. Ewing, et. al.:** Exposure to Airborne Asbestos Associated with Installing Cable Above a Suspended Ceiling. *Amer. Ind. Hyg. J.* 52(11):479–484 (1991).
15. **Keyes, D.L. and J. Chesson:** "Summary of a Study of Asbestos Exposure During Cleaning Activities." Paper presented at the National Asbestos Council Conference, San Antonio, TX (1990).
16. **Kominsky, J.R., R.W. Freyberg, and J. Chesson:** Evaluation of Two Cleaning Methods for the Removal of Asbestos Fibers from Carpet. *Am. Ind. Hyg. Assoc. J.* 51(9):500–504 (1990).
17. **U.S. EPA,** 40 CFR Part 763, Asbestos-Containing Materials in Schools, Final Rule and Notice. *Federal Register* 52(210):41825–41905.
18. **Millette, J.R., T. Kremer, and R.K. Wheeles:** "Settled Dust Analysis Used in Assessment of Buildings Containing Asbestos." *Microscope* 38: 215–220 (1990).
19. **Sebastian, P.:** "Direct or Indirect Microscope Methods for Measuring Asbestos Air Pollution in the Environment." Paper presented at the American Industrial Hygiene Conference in Orlando, FL (1990).
20. **U.S. EPA Office of Pesticides and Toxic Substances:** *Comparison of Airborne Asbestos Levels Determined by Transmission Electron Micros-*

copy (TEM) Using Direct and Indirect Transfer Techniques. U.S. EPA Pub. No. 560/5-89-004, Washington DC (1990).

21. **Yamate, G., S.C. Agarwal, and R.D. Gibbons:** *Methodology for the Measurement of Airborne Asbestos by Electron Microscopy.* U.S. EPA Office of Research and Development (1984).
22. **Gilbert, R.O.:** "Statistical Methods for Environmental Pollution Monitoring." Van Nostrand Reinhold, New York, p. 152 (1987).
23. **Miller, R.G.:** "Simultaneous Statistical Inference." Springer-Verlag, New York (1981).
24. **Chesson, J., J.D. Rench, B.D. Schultz, K.L. Milne, M. Lacker, E. Holstein, R. Lilis, and E. Crucker:** Interpretation of Airborne Asbestos Measurements. *Risk Analysis* 10(3):437–447 (1990).
25. **Michaels, D. and S. Zoloth:** "Asbestos Disease in Sheet Metal Workers: II. Radiologic Signs of Asbestos Among Active Workers." *Am. J. Ind. Med.* 12:595–603 (1987).
26. **Michaels, D. and S. Zoloth:** "Asbestos Disease in Sheet Metal Workers: Proportional Mortality Update." *Am. J. Ind. Med.* 13:731–734 (1988).
27. **Huncharek, M. and J. Muscat:** "Pleural Mesothelioma in a Non-Shipyard Electrician." *Br. J. Ind. Med.* 47:68 (1990).
28. **Guillemin, M.P., P. Madelaine, G. Litzistorf, P. Buffat, and F. Iselin:** "Asbestos in Buildings: The Difficulties of Reliable Exposure Assessment." *Aeros. Sci. Tech.* 11:221–243 (1989).
29. **Dement, J.M. and K.M. Wallingford:** "Comparison of Phase Contrast and Electron Microscope Methods for Evaluation of Occupational Asbestos Exposures." *Appl. Occup. Environ. Hyg.* 5(4):242–247 (1990).
30. **Willeke, K. and P.A. Baron:** "Sampling and Interpretation Errors in Aerosol Monitoring." *Am. Ind. Hyg. J.* 51(3):160–168 (1990).

AUTHOR AFFILIATIONS

A Diagnostic Environmental, Inc., 350 Franklin Road, Suite 300, Marietta, GA 30067

B Chesson Consulting, Inc., 1717 Massachusetts Avenue, NW, Suite 601, Washington, DC (Present Address: Resource Assessment Commission, Fisheries Resources Branch, Bureau of Resource Sciences, P.O. Box E11, Queen Victoria Terrace, Canberra ACT 2600, Australia.)

C Law Engineering, Inc., 396 Plasters Avenue, NE, Atlanta, GA 30024.

D Gobbell Hays Partners, Inc., 217 Fifth Avenue, N, Nashville, TN 37219.

E Environmental Sciences, Inc., 105 East Speedway Boulevard, Tucson, AZ 85705.

F Materials Analytical Service, Inc., 3597 Parkway Lane, Suite 250, Norcross, GA 30092.

G MVA, Inc., 5500 Oakbrook Parkway, Suite 200, Norcross, GA 30093.

Source: Environmental Choices Technical Supplement. March/April 1993. Used with permission of The Environmental Information Association, Atlanta, GA.

APPENDIX 2

DECON: A Case Study in Technology

Steve M. Hays, P.E. and James R. Millette, Ph.D.

The methods for determining the amount of asbestos in asbestos-containing building material (ACM) and for handling the removal of ACM from school buildings are well-documented under the Environmental Protection Agency (EPA) rule known as the Asbestos Hazard Emergency Response Act (AHERA),[1] and in other EPA publications.[2–3] The guidance provided in these EPA documents generally can be applied to most building situations with a few appropriate modifications. Although the EPA has pursued some work on contaminated carpets,[4] the questions from building owners confronted with items contaminated by asbestos-containing dust (ACD) only recently have begun to receive federal agency attention.[5]

To a building owner or tenant in a public building, it is important to know when furnishings, equipment and other office finishes are contaminated with asbestos; how these items can be cleaned; and what procedures are necessary during and following the cleaning process to ensure that the articles are cleaned without contaminating workers or other areas in the building.

BACKGROUND

In 1988, the architectural, engineering and asbestos management firm Gobbell Hays Partners, Inc. (GHP), and McCrone Environmental Services, Inc. (MES), a full-service asbestos analysis company, were called upon to assist a tenant in a multistory office building in answering just such questions about ACD-contaminated furnishings. Office areas occupied by the tenant comprised several floors in the upper part of a downtown high-rise building. ACD from asbestos in the building and an asbestos abatement project on a lower floor appeared to be contaminating the tenant's space. In addition, a fire occurred during the asbestos abatement project, and the resulting smoke distributed ACD and soot throughout the tenant's space.

In order to move out of the contaminated space to offices in a building that did not contain asbestos, the prospective landlord required that the tenant ensure that asbestos-contaminated furnishings would not be moved into the new space. In order to meet this requirement, an extensive decontamination program was designed, monitored and completed using state-of-the-art engineering and analytical concepts.

This was a large cleaning project; the offices contained over 4,000 desks, chairs, file cabinets, credenzas and movable wall partitions, as well as computers, telephones, typewriters and calculators. There were also objects of art and personal mementos that could not be discarded.

GHP's management approach centered on using state-of-the-art technology, as practiced by MES, to determine if surface contamination existed and to design a methodology that would adequately decontaminate the wide variety of surfaces and equipment present in the tenant space. Time was critical, since the tenant had not been able to re-enter the space after the fire to obtain needed files, equipment and furnishings to re-establish business operations.

Three basic questions had to be answered:

- Did surface contamination by asbestos fibers exist in the tenant spaces, and to what level?
- What cleaning methods and work practices would be sufficient to decontaminate the surfaces?
- What quality assurance monitoring was needed to ensure that the cleaning procedures were working?

CONTAMINATION ASSESSMENT

Standard EPA and Occupational Safety and Health Administration (OSHA) methods were utilized in the assessment of asbestos contamination in the air spaces of the office. Area air samples were collected to monitor the ambient asbestos concentrations. Personal monitors also were used to determine potential exposures of individuals.

While there has been no EPA-approved method for the analysis of ACD, MES has been working with the EPA in evaluating what has been referred to as the microvacuuming technique of dust sample collection. The protocol for collection, sample preparation and analysis is given as follows.

Samples of dust are collected using a conventional air cassette/personal pump assembly that serves as the microvacuuming tool. The cassette is moved over a premeasured area (usually one foot square) for two minutes while the pump is operating at two liters per minute. The cassette is capped, marked with identifying information and sent to the laboratory, where it is washed down with particulate-free water and refiltered onto another filter. This filter is prepared for transmission electron microscopy (TEM) analysis using the NIOSH 7402 preparation procedure. The particulate on the grids is magnified 20,000 times and asbestos fibers are identified based on their morphology, crystal structure and chemical composition. The number of asbestos fibers found in a specifically selected area on the electron microscope grid is related to the original sample area, and a value is reported in terms of fibers or structures per square foot.

After the fire, the tenant evacuated all of the firm's leased spaces (five floors) in the building and asked GHP to determine if reoccupation of the space was advisable before decontamination. First, GHP and MES personnel made an initial post-fire visual inspection of the now-unoccupied space.

After this initial visual inspection, MES began the task of collecting surface dust samples in the tenant spaces. The area air data samples taken during the initial post-fire inspection generally were low (≤0.019 structure per cubic centimeter s/cc of air by TEM); but the personal samples taken during this initial walkthrough were unacceptably high (0.066 s/cc and 0.079 s/cc by TEM). Some of the surface samples collected by MES were over 60,000,000 fibers per square foot of surface area (f/ft^2). Based on both firms' experience, this level of asbestos fibers in surface dust was very much above expected background levels. Because of the high level of surface contamination and the high personal samples (which indicated easy entrainment of surface dust into the breathing zone), the tenant decided not to reoccupy the spaces.

MES also collected surface samples from the buildings the tenant was considering for new space and which contained no ACM. The limit of detection for the surface protocol was approximately 150,000 f/ft^2; to improve the detection limit was prohibitively costly and time-consuming. All samples collected in these buildings were below this limit; therefore, the baseline for cleaning was established at 150,000 f/ft^2.

Restec Contractors, Inc., a Seattle-based abatement contracting firm, was retained by the tenant as the decontamination contractor. The firm immediately isolated the tenant's floors from the remainder of the building via critical barriers at elevators, fire exits, HVAC vents and in other areas.

Having confirmed that surface contamination existed, apparently caused by earlier abatement operations on another floor, as well as the fire and the firefighting process, GHP began to design a decontamination process. Fully contained decontamination units were erected and experiments begun to determine what cleaning methods would achieve the desired result of ≤150,000 f/ft^2.

DECONTAMINATION TECHNOLOGIES

Decontamination experiments and subsequent quality-control testing indicated that the following methods were successful in achieving desired results:

Smooth Surfaces. These surfaces included items made of metal, hardwood, smooth leather, and articles such as paper files. Objects were moved to a precleaning area that was a part of the contaminated office space area on each affected floor. The objects were vacuumed using a heavy-duty high-efficiency particulate air (HEPA) filtered vacuum, then wet-wiped using a damp cloth. Objects then were moved through an airlock into a final cleaning chamber, which was positively pressurized with clean HEPA-filtered air from a clean room that had been established in the elevator lobby on each floor previously occupied by the client. There the items were HEPA-vacuumed and wet-wiped a second time.

Objects were moved from the final cleaning chamber to a clean holding area to await quality control analytical results from randomly collected post-decontamination surface samples. Air quality was monitored in the final cleaning chamber, the holding area and the clean room.

Fabrics. Objects made with textiles, such as upholstered chairs, were treated in a precleaning area similar to the one above. In the final cleaning chamber, each item was HEPA-vacuumed again and any hard surfaces were wet-wiped. Then an industrial steam-cleaning apparatus was used to simultaneously apply a steam solution and vacuuming operation. Vacuuming was repeated several times without the steam solution. Items were allowed to dry in the clean holding area.

Electronics. Selected electronic items, such as computers circuit boards, that could not easily be replaced were HEPA-vacuumed, and the external hard surfaces selectively wet-wiped in the precleaning area. Next, these objects were moved to an intermediate cleaning chamber that was exhausted via a HEPA fan unit. Sixty psi air was directed onto surfaces from a position squarely in front of the HEPA fan inlet. Then items were HEPA-vacuumed again in the final cleaning chamber.

In general, wet-wiping practices consisted of thoroughly wiping the surface in one direction, folding the wipe cloth (or using a new one) and wiping the surface in a direction 90 degrees to the first stroke, reciprocating wipe strokes. HEPA vacuuming was performed in one direction, then in a direction 90 degrees to the first. Fabrics were best cleaned by vacuuming in a direction parallel to the warp yarn of the weave, then parallel to the weft yarn.

Quality control (QC) surface contamination samples were collected on random pieces after decontamination. The QC data, analyzed by TEM, indicated that the desired level of ≤ 150,000 f/ft^2 was reached by the cleaning procedures. As we discovered, once furnishings are found to be contaminated with ACD, they can be cleaned with a carefully designed decontamination process that uses state-of-the-art cleaning procedures and engineering controls.

REFERENCES

1. **Environmental Protection Agency.** 1987. Asbestos-containing materials in schools final rule and notice. *Federal Register.* Vol. 52, No. 210. Oct. 30.
2. **Office of Toxic Substances.** 1985. *Guidance for Controlling Asbestos-containing Materials in Buildings.* EPA 560/5-85-024, USEPA, Washington, DC.
3. **Office of Air and Waste Management.** 1985. *Sprayed Asbestos-containing Materials in Buildings: A Guidance Document.* EPA 450/2-78-014, USEPA, Washington, DC.
4. **Wilmoth, R.C., T.J. Powers and J.R. Millette.** 1988. "Observation on Studies Useful to Asbestos O&M Activities." NAC Conference, Atlanta, GA. Feb. 23.
5. **USEPA, Risk Reduction Engineering Laboratory.** 1989. *Sampling and Analysis of Asbestos in Settled Dusts Workshop.* Cincinnati, OH. July 11–14.

AUTHOR AFFILIATIONS

Steve M. Hays, P.E., is chairman of the board with Gobbell Hays Partners, Inc. Hays currently serves on the National Asbestos Council's (NAC) board of directors and the reciprocity task force. He received a bachelor's of engineering/chemical engineering degree from Vanderbilt University.

James R. Millette, Ph.D., is vice president and general manager of McCrone Environmental Services, Inc., of Norcross, GA. He has been involved in environmental/toxicology/asbestos studies since 1972. Millette received a master's in engineering from Miami University and a Ph.D. from the University of Cincinnati.

Source: Asbestos Issues, February, 1990.

APPENDIX 3

Asbestos at a Fire Scene: The Case of the Dupont Plaza Hotel

Steve Hays

On December 31, 1986, a fire at the Dupont Plaza Hotel in San Juan, Puerto Rico, injured more than 100 guests, killed 97 others, and extensively damaged the hotel. The next day, Puerto Rican authorities invited a team of investigators from the Bureau of Alcohol, Tobacco, and Firearms and from the NFPA to help determine the cause and origin of the deadly blaze.[1]

The investigation was brought to a sudden halt three weeks later when investigators noticed a crumbling white material in the debris. They suspected that it contained asbestos, a suspicion that was confirmed when a review of the hotel's construction plans revealed the presence of asbestos in the building's ceiling materials.

The site was closed, and an investigator recommended that someone with expertise in asbestos management and abatement design be retained immediately. A consulting firm from Nashville, Tennessee, was called in, and a representative arrived in San Juan the next evening.

The consultant accompanied attorneys and hotel representatives to the site the following morning. Those who actually entered the burned areas wore respirators and protective clothing as they collected bulk samples of the building materials, primarily the ceiling plaster, that were suspected of containing asbestos. Most of the ceilings on the first and second levels of the hotel had collapsed, and the plaster on the floor had been crumbled into a powder by all the foot traffic. Samples of the substance were sent to a laboratory for analysis using polarized light microscopy, which confirmed the asbestos content of the plaster.

A visual inspection of the plaster suggested that there was significant potential for airborne fibers because the plaster was crumbled and friable. The constant ocean breeze through the first two floors of the hotel, which had essentially been open to the atmosphere since the disaster, could easily entrain fibers into the air.

Airborne samples were taken using air sampling pumps with filter cassettes which were then analyzed by phase contrast microscopy and transmission electron microscopy. The analyses indicated a large concentration of airborne fibers, especially during any kind of disturbance such as walking—obviously the kind of activity necessary to complete a fire investigation. Some of the data indicated fiber concentrations higher than limits allowed by the Occupational Safety and Health Administration (OSHA).

When there was enough data to understand the situation, the hotel owners called a press conference to explain to reporters why the investigation had been halted and to outline the steps that would be taken to handle the situation.

THE EFFECTS OF ASBESTOS

Respiratory effects from long-term exposure to asbestos are well-documented. Asbestos-containing building materials are, and were at the time of the Dupont Plaza Hotel fire, regulated by OSHA and the Environmental Protection Agency. It is standard practice in the industry to use certain engineering techniques to control and contain asbestos fibers and to make workers exposed to airborne fibers use the proper respiratory protection and protective clothing.

The concern at this particular site was not with the adverse medical effects, such as asbestosis, that can be produced by long-term exposure to asbestos in the workplace. Rather, it was with the effects of short-term exposure to the episodic high levels of asbestos fibers that investigators would create as they proceeded with their task. Medical evidence shows that short-term exposure to the episodic high levels of asbestos fibers that investigators would create as they proceeded with their task. Medical evidence shows that short-term exposure to high levels of asbestos can be linked to serious cancers such as mesothelioma. Reopening the site therefore required that investigators be protected with proper respiratory equipment and clothing. It was not practical for an asbestos abatement contractor to clean the site because cleaning would destroy evidence critical to the investigation.

The local OSHA officials did not concur with the consultants' concern about short-term high-level exposure. Their position was based on OSHA regulations, which are intended only to prevent asbestosis caused by exposure to the material in the workplace over 20 to 30 years. The analytical techniques OSHA used could only determine airborne concentrations of large fibers, not the smaller fibers of the pulverized ceiling material present in the air at the Dupont Plaza. The OSHA investigation was further hampered because their analytical procedure could only determine total fiber concentration; it could not determine whether the fibers were actually asbestos fibers. The advanced procedure the consultants used was able to detect fibers much smaller than those OSHA detected and to determine specifically what type of fibers were in the air. The consultants felt that the potential for a hazardous situation definitely existed, and they maintained that position in the face of strong opposition.

The next issue was the selection of proper respiratory equipment for the investigators. The respirators commonly used in the asbestos abatement industry are half-mask air-purifying, full-facepiece air-purifying, powered air-purifying, full-facepiece supplied-air, and self-contained breathing apparatus.

Supplied-air respirators would have provided one of the highest levels of protection, but in this particular setting of twisted metal and other debris, the air lines would have presented the investigators with a significant safety hazard. Half-face cartridge respirators allow the greatest ease of use and comfort, but the consultant felt that they did not provide adequate protection against episodic exposure.

In the end, it was decided to use powered air-purifying respirators, in which filtered air is supplied to the respiratory facepiece via a battery-powered fan. The tight-fitting, full facepiece under positive pressure with filtered air has two major advantages. First, air is supplied mechanically, so the user's lungs are not required to do the work of pulling air through the filters. And second, positive pressure inside the mask protects the user from any leakage around the facepiece.

An important factor in making this choice was the inexperience of the individuals who would be wearing the respirators with the use of such equipment. It was essential to select a respirator that provided both comfort and protection for the inexperienced user.

The consultants ordered approximately $30,000 worth of respiratory equipment and protective clothing, which was shipped from the mainland United States and brought into Puerto Rico through Customs. They then conducted on-site training courses for all respirator users, instructing them in the use, care, and cleaning of their equipment and protective clothing. They also established a policy for managing entry to the site, and set up a decontamination unit for the investigators as they left the site. This allowed the investigation to begin again.

The investigators' most significant difficulty was the delay of approximately 10 days between the time the site was closed and the time the investigation was reopened. The protective equipment hindered them somewhat because it had to be donned before entry and doffed after decontamination procedures. It also restricted the investigators' movement somewhat and limited their comfort, problems that added more time to the investigation.

The decontamination procedures required of the investigators as they left the site consisted of removing their disposable protective coveralls and showering with soap and water while still wearing their respirators. After their bodies and the respirators were thoroughly clean, they could remove the respirator and change into street clothes. Decontamination procedures required disposal of the coveralls after each use; hence, expenditure for their purchase was high.

HANDLING CONTAMINATED EVIDENCE

The evidence the investigators were collecting presented a unique problem because it was assumed to have been contaminated by asbestos. Investigators tagged the material needed for future use or testing so it could be taken from the site and shipped to various analytical labs across the country. Each piece was decontaminated before shipment, using a combination of vacuuming techniques and wet wiping. It was then sealed in plastic bags or wrapped in plastic and taped, then sent to the appropriate destination. Chain of custody was obviously crucial, so thorough documentation of all procedures used on each piece of evidence was mandatory.

Subsequent analysis of the personal belongings left by guests in the hotel tower as they fled the fire indicated that they, too, had been contaminated with asbestos fibers carried up the tower by the smoke from the fire. All such belongings had to be decontaminated using vacuuming techniques before they could be shipped to their owners.

Since the fire at the Dupont Plaza Hotel created significant asbestos contamination on the burned floors and in the smoke-damaged tower above, there was some question as to how thoroughly the site, especially the mechanical systems and architectural finishes, should be cleaned before the building was reconstructed. It seemed appropriate to undertake a thorough analysis of the mechanical systems and the remaining architectural finishes after the fire investigation was complete to determine whether they had been contaminated by asbestos or any other hazardous substance, and whether those systems and finishes needed special cleaning before they were demolished and rebuilt.

It is evident that the investigation into the cause of the Dupont Plaza Hotel fire was begun without any thought of the possible presence of hazardous materials at the fire scene. It is also obvious that, in many major fires, investigators may receive episodic exposure to airborne asbestos. It would thus seem prudent for all fire investigators to consider an initial hazard assessment or hazardous materials survey of any investigation site by someone adequately trained in the use of proper respiratory equipment before they begin a thorough investigation into the cause of a fire.

REFERENCE

1. **Keam, T.J.**, "97 Die in Arson Fire at Dupont Plaza Hotel," *Fire Journal,* Vol. 81, No. 3 (May/June, 1987), p 74.

AUTHOR AFFILIATION

Steve Hays is a principal of Gobbell Hays Partners, Inc. of Nashville, Tennessee, the firm that conducted the asbestos management program at the Dupont Plaza Hotel.

Source: Fire Journal, March/April, 1990.

APPENDIX 4

Methods for the Analysis of Carpet Samples for Asbestos

James R. Millette, Ph.D.[A] P.J. Clark,[B]
Kim A. Brackett, Ph.D.[C] and R.K. Wheeles[A]

KEY WORDS

Asbestos, Transmission Electron Microscopy (TEM), Carpeting, Ultrasonic Treatment, Microvacuum

Assessing asbestos fiber contamination in a carpet is complicated by the nature of the carpeting—because of the pile's rough surface and thickness, samples cannot be collected directly from carpet for analysis by TEM. Two indirect methods are currently used by laboratories when preparing samples for measuring the amount of asbestos present in carpet material. One is an ultrasonic shaking technique which requires that a portion of the carpet be cut out and sent to the laboratory. The other is a micro-vacuuming technique which has been used generally in the assessment of asbestos in settled dust in buildings. It is not destructive to the carpet. Both methods utilize TEM to identify, measure and count the asbestos fibers found. Each can provide important but different information when an assessment of the level of contamination of carpeting is being made. The ultrasonic shaking (bulk-carpet sonication) technique gives an index of the asbestos containment throughout the entire carpet piece and the micro-vacuuming technique gives an index of the readily releasable asbestos fiber from the carpet surface.

A major concern in buildings that contain asbestos-containing material (ACM) is the extent to which the carpet may serve as a reservoir of asbestos fibers that have been released from the ACM by one mechanism or another.[1]

The Asbestos Hazard Emergency Response Act (AHERA) requires that all carpet in areas of school buildings in which ACM is present be cleaned with either a high-efficiency particulate air (HEPA) filtered vacuum cleaner or a hot water extraction cleaner (steam cleaner). The potential for airborne asbestos fiber reentrainment during carpet cleaning activities was shown in studies in which airborne asbestos concentrations were found to be between two and four times greater during cleaning than before the carpet cleaning activities.[2–3]

There are currently two methods used by the authors' laboratories for collecting and indirectly preparing samples to evaluate the amount of asbestos in or on a carpet. Direct collection and direct preparation procedures such as tape lift sampling have been investigated and cannot be used effectively with carpeting. There are currently no standard EPA methods for assessing asbestos in carpeting. One of the methods which is used is an ultrasonic extraction procedure in which a square piece cut from a carpet is mildly sonicated in water with the surfactant to release asbestos fibers both on the surface and embedded in the carpet. This method is similar to one that has been published about how to measure asbestos fibers in clothing and fabric materials.[4] The other method is a vacuuming procedure which uses a modified air sampling cassette to vacuum a sample of dust primarily from the surface of the carpet. The microvacuuming technique is non-destructive to the carpet. Both methods use the particle dispersion techniques developed over the years for the analysis of asbestos in drinking water.[5–6]

THE ULTRASONIC PREPARATION PROCEDURE

Samples of carpet are collected by cutting a piece (usually 10 centimeters by 10 centimeters) from the carpet with a razor blade or utility knife and placing it in a wide-mouth polyethylene jar or Ziploc™ bag. In the laboratory, five (5) centimeter by five (5) centimeter squares of carpet are cut and placed carpet-side down in a 1,000 milliliter beaker containing 100 milliliters of 0.1% solution of the surfactant aerosol OT or a 0.002% solution of the surfactant methyl cellulose in particle-free water. The beaker is placed in an ultrasonic bath for 30 minutes. The carpet piece is removed and rinsed into the beaker with 100 milliliters of particle-free water. The entire suspension (200 milliliters) is then shaken vigorously by hand to disperse the particles and then allowed to sit for two minutes to allow the denser particles to sink and the light particles to float to the top. At this time, three measured aliquots of different volumes (usually one (1), ten (10) and fifty (50) milliliters) are extracted with disposable graduated pipettes 1/4 to 1/2 inch below the water surface in the beaker. The aliquot is mixed with particle-free water to make 50 milliliters and filtered through a 0.22 µm pore size mixed cellulose ester filter or a 0.2 µm polycarbonate filter backed by an 0.45 µm pore size cellulose ester filter. If by visual observation, the initial beaker contains large non-asbestos fibers (carpet material), the entire suspension is passed through a coarse stainless steel mesh screen. The filters are dried and prepared for TEM analysis according to the NIOSH 7402 preparation procedure (cellulose ester filter)[7] or the Yamate, et al., procedure (polycarbonate filter).[8] At least two TEM grids from different areas of the filter are prepared for each sample. After the three filtrations are completed, the remaining suspension is transferred to a graduated cylinder and the volume recorded. This volume is added to the volumes of the measured aliquots to obtain the total volume of the sample. This accounts for the variable amount of water absorbed in the carpet during processing. A sample blank is prepared in an identical way as the sample, although no carpet segment is actually used in the ultrasonic procedure.

THE MICROVAC TECHNIQUE

Samples which are collected by microvacuuming are commonly referred to as microvac samples or microvac dust samples. They are collected by vacuuming a 100 cm^2 area (or other known area) of carpet with a membrane filter air-sampling cassette and vacuum pump. The sampling assembly consists of a 25-mm diameter mixed cellulose ester filter contained in a three-piece standard AHERA[9] air cassette with a one (1) inch piece of tubing attached to the face cap as a nozzle.[10] The end of the nozzle is cut at 45 degrees. The cassette is connected to a personal sampling pump with flexible tubing. The pump and cassette assembly are calibrated to 2 L/min. The 100-cm^2 area is vacuumed by moving the filter cassette nozzle across the carpet to agitate the carpet pile. The carpet is vacuumed for approximately 30 seconds in one direction, then another 30 seconds in a direction 90 degrees to the first. After one minute of vacuuming, the pump is turned off and the filter cassette nozzle is plugged and the cassette is labeled.

In the laboratory, the unopened microvacuuming cassettes are wet-wiped and then prepared for analysis under clean room conditions. The cassettes and filters are rinsed out with particle-free water and refiltered through a second filter which is used in the analysis. The original filter is washed during the sample preparation procedure but otherwise is not used in the preparation.

Specifically, the plug from the nozzle of the cassette is removed and the cassette is filled with approximately 10 ml of prefiltered water. The plug is replaced and the cassette is shaken vigorously by hand for two to three seconds. The entire cap of the cassette is removed and the suspension poured into a pre-cleaned 200 ml glass medical specimen bottle. All visible traces of the sample are rinsed into the specimen bottle with a plastic squirt bottle of filtered water. This procedure is repeated two additional times for a total of three washings. Next, the nozzle is rinsed two or three times into the specimen bottle. Typically, the total amount of water used in the rinse is approximately 70 to 75 ml. The water level in the specimen bottle is then carefully adjusted to 100 ml with prefiltered water. The pH of the water is adjusted to 3-4 using a 1.0% HCl solution. The sample container is capped and ultrasonicated for three minutes to make a uniform suspension. After two minutes of settling, a measured volume of suspension is extracted with a graduated pipette inserted halfway into the sample solution. The aliquot is mixed with particle-free water in the filter funnel and filtered through a 0.22 μm pore size mixed cellulose ester filter backed by an 5.0 μm pore size cellulose ester filter. The filter is dried and prepared according to the AHERA air sample preparation procedure. A sample blank is prepared in an identical way as the sample, although no carpet segment is actually vacuumed.

ASBESTOS COUNTING

In the transmission electron microscope, the number of each type of asbestos structures, chrysotile or amphibole, is determined by examining a known area on the grid in terms of a number of grid openings. The asbestos fibers are identified on the basis of morphology, selected area electron diffraction (SAED) and/or energy dispersive X-ray spectrometry (EDS).

The samples are counted following the EPA (Yamate) Provisional Method or the AHERA counting rules. The choice of counting methods depends on the particular interest of the analyst. The Yamate counting rules provide more information about the sizes of structures found by the analyst than do the AHERA counting rules. The amount of asbestos in a given sample is expressed as structures per square foot, structures per square centimeter or structures per square meter. The value is calculated using the following equation:

$$\frac{\text{EFA} \times 100 \times \text{NOSTR}}{\text{GO} \times \text{GOA} \times \text{SPL} \times \text{V}} = \text{Structures/area of the carpet}$$

NOSTR = Number of asbestos structures counted in the analysis

EFA = Effective filter area of the final sampling filter in square millimeters

GO = Number of grid openings analyzed

GOA = Average area of one TEM grid opening in square millimeters

SPL = Amount of carpet area sampled (in square feet or square centimeters)

V = Volume of sample filter

DATA ON PRECISION AND PERCENT RECOVERY

Studies on the precision and level of asbestos of the bulk-carpet piece ultrasonic shaking (sonication) technique and the micro-vacuuming technique were performed as part of the 1988 EPA study.[9] Six samples were collected using each method from carpet artificially contaminated with approximately 1 billion (1×10^9) asbestos structures per square foot (s/ft) or 1.08×10^6 str/cm^2. The artificial contamination was accomplished by spraying a known area of a carpet with a water solution of known asbestos concentration. Because there was no independent way to measure the concentration of asbestos on the carpet, no accuracy determination could be made. However, the relative efficiency of recover of one method to the other could be assessed. The mean asbestos recovery using the micro-vacuuming technique was 2.3×10^7 s/ft^2 (2.5×10^4 s/cm^2). This was approximately 3% of the mean recovery of the bulk-carpet sonication extraction technique which was 7.9×10^8 s/ft^2 (8.5×10^5 s/cm^2). The calculated coefficient of variation (CV) for the microvacuum technique was 166%. The CV for the bulk-carpet sonication procedure was 43%. It should be noted that the values given in reference 2 for structures per square foot are the correct values.[11] There was an English/metric conversion error in the article which provided incorrect values for structures per square meter throughout the paper.

A similar set of tests using "real world" carpeting that had been contaminated over a 15-year period of normal use has been performed by the authors' laboratories. The carpet samples were collected from a cafeteria in the Social Security Administration in Baltimore, Maryland. The cafeteria had an acoustical plaster ceiling containing 1 to 5% chrysotile. All furnishings had been removed from the cafeteria and the area had been vacuumed with a conventional dry vacuum cleaner twice before the samples were collected. Previous use and traffic patterns were not taken into account in collecting the samples. The samples were collected in a random manner and some samples may have been from areas where an appliance such as a soft drink machine may have stood previously. The data which include some interlaboratory comparisons are presented in Table 1. Although, there appears to be one outlier in the data set, the relative standard deviation was calculated using all eight data points for each method. The S_r for the microvacuum technique was 77% and for the bulk-carpet sonication procedure was 51% for Lab A and 47% for Lab B. The mean recovery of the microvacuuming technique compared to the bulk-carpet sonication procedure was 1%.

DISCUSSION

The answer to the question about which method is best for assessing the asbestos level in carpeting depends on the specific question being asked. In the 1988 EPA carpet study the authors concluded that sonication of bulk-carpet samples provided a more precise and accurate estimate of asbestos concentrations in carpet than the microvacuuming sampling technique. Their conclusion was based in part on data which showed that the microvacuuming technique was shown to recover significantly less asbestos from the carpet than the ultrasonic extraction technique.

The fact that one technique is more efficient in recovering fibers than the other may be important for some studies investigating the cleaning of the total carpet. However, there is reason to believe that carpets can act as a trap for asbestos fibers and that activities short of carpet removal may not disturb asbestos fibers which have worked their way deep into the carpet pile.

For those fibers which can be readily reentrained into the air from the carpet surface during cleaning activities, the microvacuuming procedure may be more appropriate. This is reasonable considering that the microvacuuming procedure collects fibers from the top layer of the carpet and the bulk-carpet sonication procedure shakes out more fibers which may be embedded deep in the carpet pile. The data show that the sonication of bulk-carpet samples is a more precise procedure than the microvacuuming procedure. However, the microvacuuming procedure has an advantage of being non-destructive. While some building owners may be willing to have a piece of carpeting cut from their building if the carpet is to be removed, it is less

Table 1. Results of carpet analyses for asbestos samples randomly chosen in cafeteria carpet after conventional (dry) vacuuming (all values in asbestos structures per square centimeter).

SAMPLE #	LAB A (SONICATION)	LAB B (SONICATION)	LAB B (MICROVAC)
1	5,400,000	4,800,000	21,000
2	3,050,000	3,300,000	30,000
3	*68,000	*5,400	*<350
4	3,600,000	3,800,000	74,000
5	3,400,000	3,000,000	50,000
6	4,300,000	2,500,000	95,000
7	3,200,000	3,600,000	18,000
8	2,000,000	4,700,000	35,000

* at detection limit

likely that a piece may be cut if the intent is to study the effectiveness of various cleaning procedures.

It is probable that a relationship exists between carpet contamination and air levels of asbestos produced during cleaning and other activities. For a given carpet, a higher level of asbestos in the carpet would be expected to produce a higher level in the air for a particular activity. The two methods described here will provide the basis for evaluating the level of asbestos contamination in carpeting to be compared with levels of asbestos in the air produced during studies of reentrainment.

REFERENCES

1. **Wilmoth, R.C., T.J. Powers and J.R. Millette,** "Observations on Studies Useful to Asbestos O&M Activities," *Microscope,* 39:299–312 (1991).
2. **Kominsky, J.R., R.W. Freyberg, J. Chesson, W.C. Cain, T.J. Powers, and R.C. Wilmoth,** "Evaluation of Two Cleaning Methods for the Removal of Asbestos Fibers from Carpet," *Am. Inc. Hyg. Assoc. J.* 51(9): 500–504 (1990).

3. **Kominsky, J.R., R.W. Freyberg, K.A. Brackett, B.A. Hollett, R.C. Wilmoth, and P.J. Clark,** "Evaluation of Three Cleaning Methods for Removal of Asbestos Fibers from Carpet and Associated Airborne Fiber Concentrations." Presented at Environmental Management '92, the Ninth Annual Convention and Exposition of the Environmental Information Association, April 8, 1992, Pittsburgh, PA.
4. **Chatfield, E.J.,** "Analytical Protocol for Determination of Asbestos Contamination of Clothing and Other Fabrics," *Microscope,* 38:221–222 (1990).
5. **Chatfield, E.J. and M.J. Dillon,** *Analytical Method for Determination of Asbestos Fibers in Water,* U.S. EPA-600/4-84-043, 1983.
6. **Brackett, K.A., P.J. Clark and J.R. Millette,** "A Method for the Analysis of Asbestos Fibers in Water Using MCE Filters." *Microscope* 40(3): 159–163 (1992).
7. **Carter, J.W., P.A. Baron and D.G. Taylor,** "NIOSH 7402." *NIOSH Manual of Analytical Methods.* pp 7401-1–7402-7 (1987).
8. **Yamate, M., S.C. Agarwal and R.D. Gibbons,** "Methodology for the Measurement of Airborne Asbestos Concentrations by Electron Microscopy." *Draft Report,* Washington, DC: Office of Research and Development, U.S. EPA, Contract No. 68-02-3266, 1984.
9. U.S. EPA, Asbestos Hazard Emergency Response Act (AHERA) Regulations, (40 CFR 763 Subpart E). October 30, 1987.
10. Microvacuuming samplers are commercially available from Environmetrics, Inc., 1019 Bankton Dr., Charleston, SC 29406.
11. Personal communication: Kominsky, J.R. and R.W. Freyberg, Environmental Quality Management, Inc., Cincinnati, OH.

AUTHOR AFFILIATIONS

A James R. Millette, PhD and R.K. Wheeles, MVA, Inc. 5500 Oakbrook Pkwy., #200, Norcross, GA 30093.

B P.J. Clark, U.S. Environmental Protection Agency Risk Reduction Engineering Laboratory, Mailstop 445, 26 West Martin Luther King Dr., Cincinnati, OH 45268.

C Kim A. Brackett, PhD, IT Corporation, 11499 Chester Road, Cincinnati, OH 45246.

Source: Environmental Choices Technical Supplement, March/April 1993. Used with permission of the Environmental Information Association, Atlanta, GA.

APPENDIX 5

Use of Observational Data and Experimental Studies in Developing Better Operations and Maintenance Plans

Steve M. Hays and James R. Millette

INTRODUCTION

The U.S. Environmental Protection Agency (EPA) publication entitled *Managing Asbestos in Place, A Building Owner's Guide to Operations and Maintenance Programs for Asbestos-Containing Materials*[1] provides information about developing Operations and Maintenance (O&M) Plans for buildings in which asbestos-containing material (ACM) is found. Although the *Green Book,* as the EPA publication is called, provides guidance for developing O&M programs, it does not give specific procedures that are required to be in a program. While all O&M Plans should contain many of the same elements, a specific O&M Plan should be developed for each building based on that building's particular situation. The *Green Book* describes some of the options which may be considered in an O&M Plan but leaves to the building owner the decisions about what specific practices should be incorporated into a building-specific Plan. Some options are more effective in minimizing asbestos exposure than others but also differ in the amount of time, effort and cost required to implement them. The process of choosing the best option can be assisted by considering observational data and experimental studies.

Observational data is that information about asbestos release or exposure collected while a particular task involving asbestos-containing material is being performed. For example, the results of an air sample collected on a person vacuuming a carpet in a building containing asbestos provides some information about the levels of airborne asbestos which occur during this activity. An experimental study is a more formally designed approach to obtain data which can be generalized and used in more than one specific situation. The experimental study requires that more information be documented about the activity and includes multiple sampling and analyses.

CHANGES IN O&M PRACTICES

It is important to note that an O&M Plan needs to be updated periodically to reflect current information. One institution's O&M Plan originally required daily carpet vacuuming with conventional vacuum cleaners to "keep the asbestos dust" down. After considering some observational data about increased asbestos levels in the air during routine carpet cleaning, the institution changed its O&M Plan to require vacuuming using a HEPA vacuum cleaner.[2–3] At a later point in time when information from an experimental study became available showing that steam cleaning was much more successful in removing asbestos from carpets than dry HEPA vacuuming, the institution changed its Plan again, this time discontinuing the HEPA vacuuming and substituting regular intervals of steam cleaning.[4] Because very few of the specific O&M tasks have been validated with actual studies, O&M Plans must be considered as dynamic documents, changing as new data are available.

The March 1978 EPA guidance document *Asbestos-Containing Materials in School Buildings,* commonly known as the *Orange Book,* addresses an "Asbestos Material Management Program." The document emphasizes that "a management system should be implemented to control maintenance, renovation, or repair work by either school personnel or contractors."[5] This includes:

1) Informing individuals that asbestos is present,
2) Training in proper work procedures to prevent damage to ACM,
3) Work order procedures and a sign-in or entry system to inform workers and other personnel that work in an area with ACM is planned,
4) Wet cleaning methods and the proper use of respirators by maintenance and custodial workers.

The March 1983 EPA document *Guidance for Controlling Asbestos-Containing Materials,* also called the *Blue Book,* presents a section for "Special Operations and Maintenance Procedures and Periodic Reassessment" as one of four possible approaches to the control of asbestos.[6] This document:

1) Uses the term "initial cleanup program" for interim cleaning measures, including:
   a) Steam cleaning or HEPA vacuuming carpets,
   b) Wet cleaning or HEPA vacuuming contaminated books and furniture,
   c) Dusting or mopping with wet or damp cloths or mops,
2) Encourages workers to wear respirators as a precautionary measure,

3) Recommends instruction in special operational procedures for custodial and maintenance personnel,
4) Gives examples of O&M activities requiring caution to avoid damaging ACM or resuspending fibers.

The June 1985 EPA document *Guidance for Controlling Asbestos-Containing Materials in Buildings,*[7] also referred to as the *Purple Book,* does the following:

1) Describes several O&M program activities including:
   a) Obtaining the cooperation of building maintenance and custodial managers,
   b) Educating building occupants and employees about ACM,
   c) Training custodial and maintenance workers in special cleaning techniques and maintenance precautions,
   d) Cleaning a building thoroughly using wet cleaning and HEPA vacuum techniques,
   e) Repeating cleaning monthly near surfacing materials or semi-annually near wrapped insulation,
   f) Taking special precautions before starting maintenance and construction work,
   g) Inspecting ACM at least twice a year for evidence of damage or deterioration,
   h) Continuing an O&M program until all ACM is removed.
2) Introduces a full chapter of guidance on establishing an O&M program,
3) Describes suggested O&M program elements for:
   a) Sprayed-on and trowelled-on surfacing materials,
   b) Pipe and boiler insulation,
   c) Other ACM
4) Introduces the terms "initial cleaning" and "periodic inspection" that appear in subsequent regulations and guidance documents.

The 1986 Occupational Safety and Health Administration (OSHA) revised document *Asbestos Regulations for the Construction Industry* (29 CFR 1926.58)[8] discusses the following:

1) "Small-scale, short duration" operations,
2) The "alteration, repair, maintenance, or renovation of structures" or portions of structures,
3) "Renovation and maintenance" activities,
4) Certain controls and work practices appropriate for O&M work (Appendix G of 29 CFR 1926.58)

These OSHA regulations are now being revised to clarify any ambiguities with the scope of O&M work, in addition to a number of other aspects of the 1986 regulations.

The October 30, 1987, Asbestos Hazard Emergency Response Act (AHERA) rules for asbestos-containing materials in schools, published in the *Federal Register* and developed by the EPA, stipulate that an O&M program must be implemented whenever friable asbestos-containing building material (ACBM) is present or assumed to be present in a school building.[9] Nonfriable materials that may be made friable must be treated as friable for O&M purposes. The AHERA program requirements include the following:

1) A written O&M plan (as part of required management plans),
2) Designation of a local education agency (LEA) person to insure the AHERA requirements are obeyed,
3) Worker protection rules, as promulgated in 40 CFR 763.121, the EPA "Worker Protection Rule,"
4) Initial cleaning and additional cleaning recommendations,
5) Procedures to be followed when O&M work disturbs friable ACBM,
6) Design of larger than small scale, short duration projects by an accredited designer,
7) Minor and major fiber release episode procedures,
8) Specific training requirements,
9) Requirement of periodic surveillance of ACBM every six months,
10) Requirement of reinspections every three years,
11) Recordkeeping requirements,
12) Recommended work practices and engineering controls for small scale, short duration O&M work (Appendix B of AHERA rules).

The January 1990 EPA three page memorandum, *Recommended Interim Guidance for Maintenance of Asbestos-Containing Floor Coverings*, addresses the following recommendation from the *Purple Book:*

> "Avoid removing, sanding, or stripping floor tiles containing asbestos. If tiles are removed, do not sand asbestos backing materials remaining on the floor."[7]

The memorandum discusses applicable guidelines for stripping wax or finish coats from asbestos-containing floors.

The July 1990 EPA guidance document *Managing Asbestos in Place* (*Green Book*) presents "basic information on how to develop and carry out high-quality O&M programs for managing asbestos in place to safeguard the health of all building occupants."[1] This document is the most comprehensive EPA document on the subject published to date, expanding on the guidelines introduced in the *Purple Book* in 1985. It discusses what is considered state-of-the-art guidance for establishing O&M programs. The important additions introduced in this document include:

1) A discussion of special work practices,
2) A description of the procedures for the following three types of O&M projects and the training levels for each,
   a) Sprayed-on or trowelled-on surfacing materials,
   b) Pipe and boiler insulation,
   c) Other ACM (wallboard, ceiling tile, floor tile, etc.),
3) A discussion of HVAC modifications for O&M work,
4) Examples of sample O&M work forms,
5) A discussion of O&M organizational charts.

Although this publication gives general guidelines for the implementation of O&M plans, it does not cover details for specific building situations. Instead, the document leaves the decision to the building owner. Several criteria for O&M plans are presented, including the effectiveness, time involved, effort, and cost for each procedure.

The November 1990 EPA revision to the asbestos National Emission Standard for Hazardous Air Pollutants (NESHAP) regulation discusses "individual nonscheduled operations," such as O&M work.[10] NESHAP specifically applies to individual nonscheduled operations where the total amount of regulated ACM to be removed or stripped in a calendar year exceeds 15 square meters (160 square feet), 80 linear meters (260 linear feet), or one cubic meter (35 cubic feet).

The 1988 modifications to the National Institute of Building Sciences (NIBS) 1986 document *Asbestos Abatement and Management in Buildings, Model Guide Specifications,*[11] include the following:

1) A set of procedures for dry removal of pipe insulation in cases where wet removal will cause damage to equipment (e.g., electrical or hot equipment),
2) A new section covering submittals for the repair of insulation and lagging,
3) Changes in wording for the OSHA labeling requirement and addition of new required AHERA labels for routine maintenance areas.

The National Institute of Building Sciences is in the process of developing an O&M work practices manual, which should be complete in the summer of 1992. Major O&M elements and their publication origins are presented in Table 1.

OBSERVATIONAL DATA

Observational data, information obtained from monitoring actual O&M activities, is useful to determine the efficacy of specific O&M activities. Common sense dictates that uncontrolled removal of asbestos-containing material will release asbestos fibers in large quantities and should be avoided. However, the extent of airborne asbestos levels which may be created during activities that disturb asbestos-containing materials in a minor way or disturb asbestos-containing dust as may happen in routine operations and maintenance work are not so obvious. Such data are important because they form a baseline of exposure by which to judge the effectiveness of various O&M practices. Table 2 presents a comparison of several cleaning techniques or combinations of cleaning techniques and their effectiveness for a number of different circumstances.[12] Table 3 presents observational data consisting of phase contrast microscopy levels generated during various activities that disturb asbestos-containing materials, including pipe insulation removal, fireproofing removal, ceiling tile disturbances, and cable disturbances.[12]

Although limited, results exist of the analysis of air samples related to operations and maintenance activities in which no particular regard was paid to the fact that asbestos was present. Some of the results can be found in a report compiled by the Health Effects Institute (HEI) - Asbestos Research Organization Literature Review Panel.[12] Other data are available from presentations and reports not cited by the HEI panel. The available data can be divided into 5 categories of activity:

1) Lifting ceiling tiles containing fireproofing debris,
2) Inspecting and working with equipment below fireproofing,
3) Pulling cable across tiles below fireproofing,
4) Cleaning of dust and debris in rooms below ACM,
5) Other O&M procedures.

Although some TEM data are available, for the majority of the monitored activities only PCM airborne asbestos results were determined.

Lifting of ceiling tiles containing debris on top caused levels of airborne asbestos averaging 5.45 fibers per cubic centimeter (f/cc) using phase contrast microscopy (PCM) in one study[13] and ranged from 0.02 to 1.4 f/cc (PCM) in another.[14]

Table 1. Origins of O&M elements: O&M guidance document comparison.

O&M ELEMENT	FIRST DISCUSSION
Continuing Inspection Program	Orange Book (1979)
Training	Orange Book (1979)
Alerting Personnel	Orange Book (1979)
Use of Respirators	Orange Book (1979)
Work Order Procedures	Orange Book (1979)
Wet Cleaning Methods	Orange Book (1979)
HEPA Vacuum Use	Orange Book (1979)
Air Sampling for O&M Work	Orange Book (1979)
Asbestos Waste Disposal	Blue Book (1983)
Initial Cleanup Program	Blue Book (1983)
Periodic Reassessments	Blue Book (1983)
Monthly Cleaning Procedures	Purple Book (1985)
Posting Caution Signs	Purple Book (1985)
Floor Tile Precautions	Purple Book (1985)
Asbestos Program Manager	Purple Book (1985)
Written O&M Plan	OSHA 29 CFR 1926.58 (1986)
Fiber Release Episode Procedures	AHERA (1987)
Recordkeeping Requirements	AHERA (1987)
Work Practices	Green Book (1990)
Three Types of O&M Projects	Green Book (1990)
HVAC System Modification	Green Book (1990)
Area Containment	Green Book (1990)
Training Levels	Green Book (1990)
Sample Forms	Green Book (1990)
Organizational Charts	Green Book (1990)

Table 2. Observational data: effectiveness of operations and maintenance cleaning procedures in reducing current levels of asbestos in air and dust.

LOCATION	TYPE OF SAMPLE	CIRCUMSTANCES	O&M CLEANING METHODS*	LEVELS		% REDUCTION	REFERENCE
				BEFORE	AFTER		
---	Air	After Fire	1,2	14.5 s/l	2.3 s/l	84%	Entek 1988
Floor	Dust	After Fire	1,2,3	1,120,181 ng/cm^2	3.24 ng/cm^2	<100%	Entek 1988-89
Furniture	Dust	After Fire	1,2	53,248 ng/cm^2	17.18 ng/cm^2	<100%	Entek 1988-89
Floor	Dust	Uncontrolled Vat Removal	1,2	380,475 s/cm^2	197 s/cm^2	<100%	Entek 1987
Carpet	Dust	Artificially Contaminated w/Chrysotile	4	62,000,000 s/ft^2	18,000,000 s/ft^2	71%	Kominsky 1989
Carpet	Dust	Artificially Contaminated w/Chrysotile	4	589,000,000 s/ft^2	196,000,000 s/ft^2	67%	Kominsky 1989
Carpet	Dust	Artificially Contaminated w/Chrysotile	1	47,000,000 s/ft^2	56,000,000 s/ft^2	NONE (net increase)	Kominsky 1989
Carpet	Dust	Artificially Contaminated w/Chrysotile	1	535,000,000 s/ft^2	447,000,000 s/ft^2	16%	Kominsky 1989

Table 2. Continued.

LOCATION	TYPE OF SAMPLE	CIRCUMSTANCES	O&M CLEANING METHODS*	LEVELS		% REDUCTION	REFERENCE
				BEFORE	AFTER		
Smooth Surfaces	Dust	After Fire	1,2,1,2	60,000,000 s/ft^2	≤ 150,000 s/ft^2	<100%	Hays & Millette 1990[19]
Fabric	Dust	After Fire	1,4	60,000,000 s/ft^2	≤ 150,000 s/ft^2	<100%	Hays & Millette 1990

Key

1 - HEPA Vacuum
2 - Wet Wipe
3 - Rubber Sponge Wiping
4 - HEPA Filtered Hot Water Extraction

s/l - structures per liter
ng/cm^2 - nanograms per square centimeter
s/ft^2 - structures per square foot

Table 3. Observational data: phase-contrast microscopy airborne levels produced during activities that disturb asbestos-containing materials.

ITEM REMOVED	CONTROL METHOD	AIRBORNE LEVELS (f/ml)*	NUMBER OF SAMPLES	REFERENCE
Pipe Insulation	None	4.1 (AM)	Unknown	Sawyer 1987
Pipe Insulation	Wet Removal	0.05 (M)	1649	Piper 1989
Pipe Insulation	Wet Removal	0.09 (GM)	11	Consad 1984
Pipe Insulation	Wet Removal	0.02 (PM)	38	Entek 1990
Pipe Insulation	Wet Removal	0.13 (GM)	4	Consad 1984
Pipe Insulation	Glove Bag	0.0217 (M)	20	Hygienetics 1990
Pipe Insulation	Glove Bag	0.0246 (M)	10	Kominsky & Freyburg 1989
Pipe Insulation	Glove Bag	0.04 (M)	434	Piper 1989
Fireproofing	None	74.36 (AM)	6	Sawyer 1976, 1977
Fireproofing	None	4.5 (GM)	71	Consad 1984
Fireproofing	None	16.4 (GM)	79	Paik 1983
Fireproofing	Wet Removal	0.5 (GM)	15	Paik 1983
Fireproofing (Small Area)	Wet Removal	0.1138 (M)	22	Entek 1986
Ceiling Tiles (Disturbed)	None	5.45 (AM)	2	Hamilton 1980
Ceiling Tiles (Disturbed)	None	0.14 (GM)	11	Consad 1984
Ceiling Tiles (Disturbed)	LVHV Collection w/HEPA Vacuum	0.0085 (GM)	11	Entek 1987
Ceiling Tiles (Disturbed)	LVHV Collection w/HEPA Vacuum	0.02 (AM)	11	Entek 1987

Table 3. Continued.

ITEM REPLACED	CONTROL METHOD	AIRBORNE LEVELS (f/ml)*	NUMBER OF SAMPLES	REFERENCE
Cables above Ceiling w/Sprayed Fireproofing	None	0.93 (AM)	4	Hamilton 1980
Cables above Ceiling w/Sprayed Fireproofing	None	1.1 (AM)	13	Environ- mental Sciences
Cables above Ceiling w/Sprayed Fireproofing	Pacification (Varies)	0.01	20	Hygienetics 1990
Cables above Ceiling w/Sprayed Fireproofing	Pacification (Varies)	0.05	9	Hygienetics 1990

Key M = MEAN; GM = GEOMETRIC MEAN; AM = ARITHMETIC MEAN; M = POOLED MEAN

Workers involved in inspecting or adjusting equipment above the ceiling tile produced levels of airborne asbestos averaging 1.3 f/cc (PCM)[13] and ranging from 0.04 to 0.9 f/cc (PCM).[14] When installation and repair of HVAC systems in the vicinity of insulation was included in the activity the levels ranged from 0.11 to 6.9 f/cc (PCM).

Installing cable above ceiling tiles involves disturbing asbestos-containing dust and possibly some incidental contact with the fireproofing. Data from two studies show that this type of activity results in airborne levels of asbestos averaging 0.13 and 0.34 f/cc (PCM) and 26.9 and 67.4 structures per cubic centimeter (s/cc) using transmission electron microscopy (TEM).[15] A similar study performed in an office building with a different type of fireproofing showed average area samples collected during the cable installation of 3.5 s/cc and personal samples averaging 26 s/cc by TEM.[16]

An early published study of work in warehouses where fireproofing was present showed levels of airborne asbestos ranging from 0.3 to 52.6 f/cc by light microscopy.[17] More recent data determined using TEM analysis showed levels of 29.7 s/cc for activities disturbing settled dust originating from the fireproofing in a warehouse.

Cleaning of fireproofing debris and dust was the subject of the study at the Art and Architecture building at Yale University. Custodial sweeping using dry methods resulted in airborne fiber levels of 1.63 f/cc by PCM and for dry dusting procedures levels of 4.02 f/cc by PCM.[18]

Dry sweeping of fireproofing dust after a cable pulling exercise in an office building showed airborne levels of 54 s/cc in general area sampling with personal samples averaging 61 s/cc by TEM.[16]

Some data exist which support the generally specified O&M cleaning practice of wetting the material or wet wiping the dust, and the use of low volume, high velocity (LVHV) collection of debris and dust with a HEPA vacuum for reducing airborne asbestos levels when dealing with a minor disturbance of ACM or in activities involving the resuspension of asbestos-containing dust into the air. However, the practice does not completely eliminate the exposure.

Removal of a small area of fireproofing for installation of a sprinkler system was done under O&M procedures including LVHV collection with HEPA vacuum and misting the fireproofing. The work caused levels of airborne asbestos ranging from 0.013 to 0.46 f/cc (PCM).

In a somewhat similar situation of minor disturbance of fireproofing, the collection of 6 bulk samples above suspended ceilings on 3 floors in which the collector used procedures to minimize the release of asbestos from the in-place material but not from the possible asbestos-containing dust caused levels of airborne asbestos collected on a personal monitor of 0.0096 f/cc (PCM) and 0.088 s/cc (TEM).

With O&M procedures including LVHV collection with HEPA vacuum, the airborne asbestos levels during a project involving lifting ceiling tiles ranged from 0 to 0.01 f/cc (PCM).

Monitoring of the rate of the accumulation of asbestos fibers on horizontal surfaces such as ceiling tiles below fireproofing is helpful in determining how often cleaning procedures need protective measures. The monitoring is done using passive monitoring collectors modeled after techniques used for years in outdoor dust collection studies. In buildings, forensic evidence tins are placed on horizontal surfaces such as ceiling tiles and left for extended periods of time usually in terms of months. The samples are collected and analyzed by TEM. In one set of samples collected in a school building, both passive monitor tins collected after 14 days and over 70 days showed accumulation rates of about 200 asbestos structures per square centimeter per day (s/cm^2/day). In another project, tins set out in a boiler room below fireproofing showed accumulation of about 600 s/cm^2/day.

EXPERIMENTAL DATA

Experimental data is data obtained from staging or simulating common O&M activities and taking air and dust samples to understand the levels of contamination generated by various activities. Although not as extensive as observational data, experimental data does exist for a number of different O&M activities and warrants investigation. Tables 4 and 5 show results of two different studies of airborne asbestos concentrations generated by a simulated cable installation.

In one study which involved a simulated cable pull across tiles containing asbestos dust, the air concentration detected by TEM before the experiment ranged from 0.025 to 0.097 s/cc, with a mean value of 0.052 s/cc. The levels detected during the experiment ranged from 1.5 to 46.6 s/cc, with a mean of 23.5 s/cc (TEM). Finally, airborne levels measured after the simulation averaged 6.8 s/cc, ranging from 1.1 s/cc to 21.0 s/cc (TEM). In a related experiment also simulating a cable pull, the values of airborne asbestos before the cable pull ranged from below detection to 0.032 s/cc (TEM). During the cable pull, the airborne asbestos levels ranged from 2.2 to 4.5 s/cc, with a mean value of 3.6 s/cc. After the cable pull, levels measured ranged from 1.9 to 6.2 s/cc, with a mean of 3.8 s/cc.

Another experiment examined the asbestos contamination resulting from the movement of a wall. Levels of total airborne asbestos structures measured before the activity were 0.02 s/cc (TEM), while levels detected during the wall movement were substantially higher at 68.97 s/cc (TEM). Levels measured after the experiment were also elevated above initial levels at 5.29 s/cc.

CONCLUSIONS

Based on the observational and experimental data presented, the following conclusions can be made:

1) The most effective cleaning techniques for floors and furniture ranked from most effective to least effective are:

    a) combination of HEPA vacuum, wet wipe, rubber sponge wipe (most effective);
    b) combination of HEPA vacuum, wet wipe;
    c) HEPA vacuum only (least effective).

2) The most effective control method for asbestos removal in an O&M situation, ranked from most effective to least effective, are:

Table 4. Experimental data: first study of airborne asbestos concentrations generated by simulated cable installations.

ACTIVITY	PART OF EXPERIMENT	AIR CONCENTRATION (TEM, s/cc)
Cable pull above ceiling tiles	Before	0.066
	Before	0.033
	Before	0.097
	Before	0.037
	Before	0.025 Before: Mean 0.052 Range: 0.025-0.097
	During	46.6
	During	17.0
	During	20.0
	During	24.1
	During	33.5
	During	21.7
	During	1.5 During: Mean 23.5 Range: 1.5-46.6
	After	21.0
	After	3.2
	After	7.9
	After	1.7
	After	2.9
	After	1.1
	After	11.0
	After	5.8 After: Mean 6.8 Range: 1.1-21.0

Table 5. Experimental data: second study of airborne asbestos concentrations generated by simulated cable installations.

ACTIVITY	PART OF EXPERIMENT	AIR CONCENTRATION (TEM, s/cc)
Cable pull above ceiling tiles	Before	< 0.016
	Before	< 0.016
	Before	< 0.016
	Before	0.032
	Before	< 0.016 Before: Mean < 0.019 Range: < 0.016-0.032
	During	3.9
	During	3.6
	During	2.2
	During	4.5
	During	3.6 During: Mean 3.6 Range: 2.2-4.5
	After	5.1
	After	2.0
	After	1.9
	After	4.0
	After	6.2 After: Mean 3.8 Range: 1.9-6.2

a) glove bag or LVHV collection with HEPA vacuum (most effective);
b) wet removal
c) dry removal (no control) (least effective).

3) O&M activities which tend to generate the highest levels of airborne asbestos contamination, ranked from highest to lowest, are:

   a) fireproofing removal (highest);
   b) pipe insulation removal;
   c) ceiling tile or cable disturbance (lowest).
4) High levels of airborne asbestos occur during the activity but actually may be higher during cleanup.

As these data indicate, different O&M procedures produce different results. More experimental studies are needed to determine the relative importance of the factors involved. The results depend on the type of O&M activity, the kind of ACM or asbestos contamination involved, and training and ability of the workers. The building manager should establish criteria for the O&M program (which may be those set by regulations, such as OSHA, or which may be more precautionary), then use experimental and observational data to prove the efficacy of selected procedures. Adjustments to procedures should be made as the data indicates.

REFERENCES

1. **USEPA.** 1990. U.S. Environmental Protection Agency. *Managing Asbestos in Place: A Building Owner's Guide to Operations and Maintenance Programs for Asbestos Containing Materials.* Washington DC: USEPA. EPA 20T-2003. ("Green Book")
2. **Millette, J.R.** *Dust Analysis in Buildings.* Paper presented at the National Asbestos Council Conference, Atlanta, GA, February 19, 1988.
3. **Wilmoth, R.C., Powers, T.J., and Millette, J.R.** Observations on Studies Useful to Asbestos O&M Activities. *The Microscope.* 39:299-312 (1991).
4. **Kominsky, J.R., et al.** Evaluation of Two Cleaning Methods for the Removal of Asbestos Fibers from Carpet. *American Industrial Hygiene Association Journal.* 51(9):500-504 (1990).
5. **USEPA.** 1978. U.S. Environmental Protection Agency. *Sprayed Asbestos-Containing Materials in Buildings: A Guidance Document.* Washington, DC: USEPA. EPA 450/2-78-014. ("Orange Book")
6. **USEPA.** 1983. U.S. Environmental Protection Agency. *Guidance for Controlling Friable Asbestos-Containing Materials in Buildings.* Washington, DC: USEPA. EPA 560/5-83-002. ("Blue Book")
7. **USEPA.** 1985. U.S. Environmental Protection Agency. *Guidance for Controlling Asbestos-Containing Materials in Buildings.* Washington, DC: USEPA. EPA 560/5-85-024. ("Purple Book")
8. **U.S. Department of Labor: OSHA Regulations.** 29 CFR 1926.58 - *Construction Industry Asbestos Standard.* June 1986; Amended, September, 1988.

9. **USEPA.** 1987. U.S. Environmental Protection Agency. *Asbestos-Containing Materials in Schools; Final Rule and Notice.* 40 CFR 763. *Federal Register,* October 30, 1987.
10. **USEPA.** 1990. U.S. Environmental Protection Agency. National Emission Standards for Hazardous Air Pollutants; *Asbestos NESHAP Revision; Final Rule.* 40 CFR 61. *Federal Register,* November 20, 1990.
11. **National Institute of Building Sciences.** *Asbestos Abatement & Management in Buildings, Model Guide Specifications.* Second Edition. August 12, 1988.
12. **Health Effects Institute - Asbestos Research (HEI-AR).** *Asbestos in Public and Commercial Buildings: A Literature Review and Synthesis of Current Knowledge.* 1991.
13. **Hamilton, D.J.** 1980. As cited in HEI-AR *Asbestos in Public and Commercial Buildings.* 1991. Page 5-57.
14. **CONSAD Research Corporation.** 1984. As cited in HEI-AR *Asbestos in Public and Commercial Buildings.* 1991. Page 5-55.
15. **Keyes, Dale L., et al.** Exposure to Airborne Asbestos Associated with Simulated Cable Installation Above a Suspended Ceiling. *American Industrial Hygiene Association Journal.* 52(11):479-484 (1991).
16. **Ewing, W.M., et al.** *Asbestos Exposure During and Following Cable Installation in the Vicinity of Fireproofing.* Paper presented at the American Industrial Hygiene Conference, 1991.
17. **Lumley, K.P.S., et al.** Buildings Insulated with Spray Asbestos: A Potential Hazard. *Ann. Occup. Hyg.* 14:255-257 (1971).
18. **Sawyer, R.N.** Asbestos Exposure in a Yale Building: Analysis and Resolution. *Environ. Res.* 13(1):146-163 (1977).
19. **Hays, S.M. and Millette, J.R.** Decon: A Case Study in Technology. *Asbestos Issues* 3(2):42-46 (1990).

NOTES

Session 17, Paper Number II, April 8, 1992. Environmental Management '92. The Ninth Annual Conference and Exposition of NAC/The Environmental Information Association.

AUTHOR AFFILIATIONS

Steve M. Hays, P.E., C.I.H., Chairman of the Board, Gobbell Hays Partners, Inc., 217 Fifth Avenue North, Nashville, Tennessee 37219, (615)254-8500

James R. Millette, Ph.D., Executive Director, Millette, Vander Wood & Associates, Inc., 5500 Oakbrook Parkway, Suite 200, Norcross, Georgia 30093, (404)662-8509

APPENDIX 6

Air and Dust Sampling in Return Air Plenums

Steve M. Hays

INTRODUCTION

In the asbestos detection and control industry, it is normal to find asbestos structures in the dust on tops of ceiling tiles which are suspended beneath spray-applied asbestos-containing (AC) fireproofing. Research indicates that such a condition is of concern to maintenance workers who disturb the ceiling tiles and has serious implications for the proper design and administration of an operations and maintenance (O and M) plan.[1] The spaces created by suspended ceilings and the structural decks above are often used as plenums for return air to a building's heating and air conditioning systems.

In the author's experience, standard passive air sampling techniques, with analysis by transmission electron microscopy (TEM), generally detect only low levels (0.05 s/cm^3 or less) of airborne asbestos in return air plenums under steady state conditions and absent any disturbance of plenum dust or AC materials. These low airborne concentrations often exist, under the conditions just described, even when the counts of asbestos structures in dust are substantial (1000 s/cm^2 and higher). This paper will present and discuss, with assumptions and example calculations, the hypothesis that deposition rates, expressed in asbestos structures [as defined in the United States Environmental Protection Agency (USEPA) regulations for the Asbestos Hazard Emergency Response Act] per square centimeter per minute (s/cm^2 min), are such that large counts of asbestos structures can accumulate in plenum dust while measured airborne concentrations in the plenum remain generally low. A theoretical relationship between deposition rates and airborne concentrations will be presented.

BACKGROUND

Sampling and analysis methodologies for airborne asbestos are well established in the industry.[2–4] Sampling and analysis protocols to quantify asbestos in dust, using a microvacuuming collection technique and TEM analysis, are under development by the USEPA[5] and the American Society for Testing and Materials (ASTM).[6] Methods to determine the rate at which asbestos fibers settle onto a surface have been in use for several years. One such method is in situ dust sampling, using forensic evidence tins as collection devices.[7–8] This method collects dust over a known time interval, then analyzes the dust by TEM for asbestos.

Operations and maintenance procedures are intended to prevent and/or control the release of asbestos structures into the air of a facility.[9] Data of various sorts are useful in designing procedures that accomplish this task in any particular facility of interest.[10] It is the author's opinion that plenum dust data is more important than passive, steady state plenum air data when designing an O and M plan and when designing other asbestos abatement procedures related to the plenum. The presentation and discussion of this paper's hypothesis will support this opinion. In situ dust sampling data will be used in example calculations to illustrate application of a theoretical relationship between deposition rates and airborne concentrations. Some factors which may affect air sampling results will be discussed in relation to the hypothesis.

OBJECTIVES

The objectives of this paper are:

1. To present and discuss a theoretical relationship between deposition rates of asbestos structures in plenum dust and asbestos concentrations in plenum air.
2. To use this theoretical relationship and other information to support the hypothesis that large counts of asbestos structures can accumulate in plenum dust while measured airborne concentrations in the plenum generally remain low.
3. To draw conclusions about what types of sampling and analysis may be useful in designing operations and maintenance plans.

DISCUSSION

The following assumptions are made:

1. Steady state conditions exist in the plenum. No mechanical disturbance of the AC fireproofing or the surfaces of interest occurs.

2. The deposition rate of asbestos structures on the surfaces of interest (tops of ceiling tiles, tops of light fixtures, etc.) is equal to the rate of release of asbestos structures from the source. This follows from the assumption of steady state conditions.
3. Asbestos-containing fireproofing on structural steel and decking is present in the plenum and is the only source of asbestos in the facility.

Based on these assumptions, a relationship between deposition rate and airborne concentration can be developed. For this development, theoretical settling times for asbestos in still air will be used.[11] The validity of combining theoretical settling rates for still air with empirical asbestos deposition rates to describe airborne concentrations in a return plenum, where air is not always still, will be discussed later.

The settling time t is the total time required for a structure to settle onto a surface of interest after it is released from a source, assumed in this case to be the asbestos-containing fireproofing. From the literature, the settling velocity f can be estimated from theoretical calculations for asbestos structures of approximately the same size as those found in plenum dust. The settling time t for a distance H is calculated by the formula:

$$t* = \frac{H}{f} \qquad (1)$$

*See Appendix A for units

R, the number of structures released per unit time, from a given area A of fireproofing is calculated from D_a, the measured deposition rate per unit area, by the equation:

$$R = D_a A \qquad (2)$$

During the settling time t structures released from the source are airborne in the volume V of plenum air beneath the area A of the source. The total number of structures S_a released from area A of the source into the air volume V during time t may be calculated by the formula:

$$tR = S_a \qquad (3)$$

The concentration C_v of airborne asbestos structures (the number of structures S_a per unit volume of plenum air V) can be calculated by the equation:

$$C_v = \frac{S_a}{V} \tag{4}$$

Substituting equation (3) into equation (4) yields a variation of the formula for the concentration C_v of airborne asbestos structures:

$$C_v = \frac{tR}{V} \tag{5}$$

Further substitution of equation (2) into equation (5) produces:

$$C_v = \frac{tD_aA}{V} \tag{6}$$

Since volume V can be expressed as area A multiplied by height H:

$$A = \frac{V}{H} \tag{7}$$

Substitution of equation (7) into equation (6) produces:

$$C_v = \frac{tD_aV}{HV} \tag{8}$$

Elimination of the volume V terms generates the formula:

$$C_v = \frac{tD_a}{H} \tag{9}$$

From equation (1), t = H/f, substitution into equation 9 produces the final equation for the concentration C_v of airborne asbestos structures:

$$C_v = \frac{D_a}{f} \tag{10}$$

Application of this formula is illustrated using values of D_a, the deposition rate per unit area, measured in plenum spaces of a commercial office building and values of f, the settling velocity.[11] These deposition rates were measured by placing an opened forensic evidence tin in the plenum space for 113 days. The dust collected in the tin over this time period was analyzed for asbestos structures by TEM.

The first example will use the average value of D_a for the data set from the building of:

$$D_a = 5.3 \times 10^{-4} \text{ s/cm}^2 \text{ sec}$$

For this data set, the average dimensions for an asbestos structure is estimated to be 5.0µm in length and 0.4µm in diameter, yielding a theoretical settling velocity[11] of:

$$f = 4 \times 10^{-3} \text{ cm/sec}$$

Therefore, using equation (10), $C_v = D_a/f$, the theoretical value of C_v, the concentration of airborne asbestos structures, for this example is:

$$C_v = 0.13 \text{ s/cm}^3$$

One of the lower values of D_a at this facility was 2.5×10^{-4} s/cm^{2}sec. Based on estimated average dimensions for asbestos structures from the data and the literature settling velocity[11], the theoretical value of C_v for this D_a is 0.063 s/cm^3.

The validity of this approach to estimating C_v is dependent on the validity of the underlying assumptions, the accuracy of the measured values of D_a, and the applicability of the literature settling velocities to the situation. The concentration given by this formula is highly sensitive to the settling velocity, which is theoretically dependent on fiber dimensions. Table I illustrates various values of C_v for different D_a and f values. (Table I does not represent measured values of D_a, but is presented to show how C_v will vary for possible values of D_a.)

As Table 1 demonstrates, values of C_v can be quite low. If C_v as calculated from measured values of D_a for a facility is accepted as the actual average airborne concentration, it is to be expected that measured air concentrations would be some value less than C_v. The alteration in particle trajectory caused by an electrostatic field around the sampling cassette can result in both nonuniform deposits on the filter and a net loss of sample.[12] Factors which affect this phenomenon include particle charge, sampler charge, sampler conductivity, and sampling flow rate and direction[12]. The effects of particle charge can also be magnified by low humidity levels, especially levels below 15% relative humidity.[12] Nonconductive air samplers present problems since pockets of charge may develop on the surface (due to the nonconductive nature of the sampler material), resulting in nonuniform deposits on the filter or net loss of the sample.[12] Given then that calculated values of C_v are often low and that some net loss of sample occurs using accepted air sampling protocols, it is not surprising that steady state plenum air concentrations are usually measured to be quite low. This happens even when large counts of asbestos in dust are accumulating over long periods of time.

The validity of using theoretical settling rates for still air with empirical deposition rates requires discussion. Air in return plenums is certainly not always still. Some comfort air systems cycle on and off according to the demand placed on them and some operate continuously during hours of building occupancy. It is common practice to modify the mode of operation over nights and weekends, including complete shutdowns. Deposition rates in plenums are usually measured over long periods of time: weeks or months. The value of D_a for a given plenum would reflect periods of still and moving air. (Air velocities in return plenums normally range from still to 380 cm/sec at the air grille.) Therefore, D_a represents an average deposition rate over the range of operational modes for a system during the time period of interest. Settling times t are affected by air movement[11] and would depend in a given plenum on air velocities, asbestos structure characteristics, and air flow patterns. Measuring t in a plenum would be difficult and costly. Absent this data or research data that could be suitably interpolated, it is reasonable to use literature values of t (or f) for still air. D_a is probably not constant over short periods of time, such as over an 8 hour air

Table 1. Theoretical airborne asbestos structure concentrations (s/cm^2) for varying f's and D_a's

Settling Velocity, f(cm/sec)	1×10^{-6}	5×10^{-6}	1×10^{-5}	5×10^{-5}	1×10^{-4}	5×10^{-4}	1×10^{-3}	5×10^{-3}	1×10^{-2}
1×10^{-4}	0.01	0.05	0.1	0.5	1	5	10	50	100
5×10^{-4}	2E-3	0.01	0.02	0.1	0.2	1	2	10	20
1×10^{-3}	1E-3	5E-3	0.01	0.05	0.1	0.5	1	5	10
5×10^{-3}	2E-4	1E-3	2E-3	0.01	0.02	0.1	0.2	1	2
1×10^{-2}	1E-4	5E-4	1E-3	5E-3	0.01	0.05	0.1	0.5	1
5×10^{-2}	2E-5	1E-4	2E-4	1E-3	2E-3	0.01	0.02	0.1	0.2
1×10^{-1}	1E-5	5E-5	1E-4	5E-4	1E-3	5E-3	0.01	0.05	0.1
5×10^{-1}	2E-6	1E-5	2E-5	1E-4	2E-4	1E-3	2E-3	0.01	0.02
1×10^{0}	1E-6	5E-6	1E-5	5E-5	1E-4	5E-4	1E-3	5E-3	0.01

sampling time interval; therefore, air concentrations calculated from D_a would not be expected to correlate well with individual measured air concentrations in a given plenum. However, since D_a is an average over a long time period, C_v is arguably a useful estimate of the theoretical average airborne concentration.

CONCLUSIONS

1. The theoretical average airborne concentration of asbestos structures in a return air plenum can be estimated using measured deposition rates and literature values of asbestos structure settling velocities. The relationship is:

$$C_v = \frac{D_a}{f}$$

2. Example calculations using the relationship expressed above demonstrate that large counts of asbestos structures can accumulate on surfaces in plenums while measured air concentrations can be low.
3. Airborne concentrations of asbestos structures in plenums often measure low, asbestos counts in dust can concomitantly accumulate to large quantities, and asbestos structures in dust can become airborne when the dust is disturbed by maintenance activities; therefore, quantification of asbestos in plenum dust is more useful data than plenum air concentration data when designing an O and M plan. The asbestos existing in plenum dust can be quickly quantified using a microvacuuming collection technique and the rate of asbestos deposition can be quantified with in situ dust sampling. These values are useful in determining what special procedures may be needed, if any, when performing maintenance functions in a plenum space.

ISSUES FOR FURTHER STUDY

1. The application of the relationship between deposition rate and airborne concentration could be improved by a rigorous investigation of the underlying assumptions.
2. Research to better characterize structure settling velocities in plenums over the range of normal operational modes would allow better application of the relationship.

3. Investigation of the effect on in situ dust sampling of collection tin placement in relation to air movement patterns and characteristics would allow refinement of the relationship.
4. Simultaneous collection of enough air concentration and in situ dust data would allow development of empirical relationships between C_v and D_a.
5. The effect of circulation of asbestos structures through the plenum from other asbestos-containing sources and from entrainment into the air system of asbestos-containing dust from outside the plenum should be considered.

APPENDIX A

Nomenclature

f = settling velocity in centimeters per second (cm/sec)

t = settling time in seconds (s)

R = number of structures released per unit time in structures per second (s/sec)

D_a = deposition rate per unit area per unit time in structures per square centimeter per second (s/cm^{2}sec)

S_a = total number of structures (s) released into the air from area A of a source

C_v = concentration of airborne structures per cubic centimeter (s/cm^3)

V = volume of plenum space in cubic centimeters (cm^3)

A = area of source in square centimeters (cm^2)

H = height of plenum space in centimeters (cm)

s/cm^2 = structures per square centimeter

s/cm^{2}min = structures per square centimeter per minute

REFERENCES

1. **Keyes, Dale L., et al:** Exposure to Airborne Asbestos Associated with Simulated Cable Installation Above a Suspended Ceiling. *American Industrial Hygiene Association Journal.* 52(11):479-484 (1991).
2. **Keyes, Dale L. and Chesson, Jean.** 1989. A Guide to Monitoring Airborne Asbestos in Buildings. Environmental Sciences, Inc., 105 E. Speedway Blvd., Tucson, Arizona 85705.
3. **USEPA.** 1987. U.S. Environmental Protection Agency. Asbestos-Containing Materials in Schools; Final Rule and Notice. 40 CFR 763. Federal Register, October 30, 1987.
4. **National Institute for Occupational Safety and Health**: NIOSH 7400 Method Protocol. Washington, DC: National Institute for Occupational Safety and Health, 1984.
5. **USEPA.** 1990. U.S. Environmental Protection Agency. Draft Test Method for Sampling and Analysis of Dust for Asbestos Structures by Transmission Electron Microscopy.
6. **Stiers, David.** Chair: ASTM Subcommittee D22.05.
7. **Segrave, Alan M.** In Situ Dust Sampling Protocol. *Asbestos Issues.* August 1990. p. 42.
8. **Millette, J.R.** 1992. Timed Passive Dust Collection. Millette, Vander Wood & Associates, 5500 Oakbrook Parkway, #200, Norcross, Georgia 30093.
9. **USEPA.** 1990. U.S. Environmental Protection Agency. Managing Asbestos in Place: A Building Owner's Guide to Operations and Maintenance Programs for Asbestos Containing Materials. Washington, DC: USEPA. ("Green Book")
10. **Hays, Steve M. and Millette, James.** 1992. Use of Observational Data and Experimental Studies in Developing Better Operations and Maintenance Plans. Prepared for presentation at the NAC/Environmental Information Association Conference, Pittsburgh, Pennsylvania, April 1992.
11. **USEPA.** 1978. U.S. Environmental Protection Agency. Sprayed Asbestos-Containing Materials in Buildings: A Guidance Document. Washington, DC: USEPA. EPA 450/2-78-014. ("Orange Book")
12. **Baron, Paul A. and Deye, Gregory J.** Electrostatic Effects in Asbestos Sampling I: Experimental Measurements. *American Industrial Hygiene Association Journal.* 51(2):51-62 (1990).

ACKNOWLEDGMENT

The author appreciates the assistance of Steven E. Koons in the preparation of this paper.

Session 14, Paper Number II, April 8, 1992, Environmental Management '92, The Ninth Annual Conference and Exposition of NAC/The Environmental Information Association.

AUTHOR AFFILIATION

Steve M. Hays, P.E., C.I.H., Chairman of the Board, Gobbell Hays Partners, Inc., Architects, Engineers, Environmental Consultants, 217 Fifth Avenue North, Nashville, Tennessee 37219, 615-254-8500, Fax 615-256-3439; 101 California Street, Suite 1095, San Francisco, California 94111, 415-398-7254, Fax 415-398-7184.

APPENDIX 7

Observations on Studies Useful to Asbestos O&M* Activities

Roger C. Wilmoth, Thomas J. Powers,** and
James R. Millette***

KEY WORDS

Asbestos, dust analysis, airborne dust, house dust, air sampling, phase contrast microscopy, TEM, electron microscopy.

ABSTRACT

Asbestos-containing materials found in buildings may release asbestos fibers into the air. Some of these fibers will eventually settle and attach to room surfaces (walls, furnishings, equipment, floors and carpet) as part of normal dust. Activities like dusting, sweeping and vacuuming are likely to re-entrain the dust causing exposure to airborne asbestos. This paper discusses data largely observational in nature but illustrative of general trends of interest to those individuals dealing with the day-to-day problems of asbestos in buildings.

INTRODUCTION/RISKS

Buildings that contain ACM (asbestos-containing materials) may present unique problems for custodial workers. The release of asbestos by aged and deteriorated ACM is known to be episodic, relating to a myriad of factors such as activity within the area, vibration, temperature, humidity, airflow, use patterns, etc. Hopefully, for the majority of buildings containing ACM fiber, release is minimal; however, since we are unable at this time to make accurate predictions on this, prudent judgment should prevail for the protection of building occupants and workers. Therefore, the assumption is made that special precautions must be taken for buildings containing asbestos. This conservative approach is necessary because asbestos is a known human carcinogen. In the absence of contradictory data, health scientists[1] feel that the carcinogenicity of asbestos follows a linear, zero-threshold model, which means that:

1) inhaling a single fiber represents a risk
2) doubling the exposure doubles the risk, and so on.

The inhalation of asbestos can result in the following diseases:

a) asbestosis—a debilitating lung disease similar to silicosis,
b) lung cancer—a usually fatal lung malignancy, and/or
c) mesothelioma—an incurable, fatal cancer of the mesothelial lining of the chest.

The absolute risks have been difficult to define because of the latency period (time between exposure and onset of symptoms), which for these asbestos-related diseases may be twenty or more years. It is known that smoking increases these risks by at least a factor of ten. EPA[2] estimates that between 3000 and 13000 deaths each year in the US are attributable to asbestos.

If a building contains asbestos and releases occur, the custodial worker may be most intensely affected over the long term. If a release occurs, much of the asbestos will eventually settle and attach to the surfaces of the rooms (walls, furnishings, equipment, floors, and carpet). As the custodial workers repeatedly cleans and dusts these rooms, the energy imparted by these activities may re-entrain the asbestos into the air, resulting in potentially chronic long-term exposures. Activities like dusting, sweeping, and vacuuming are mechanically aggressive and are most likely to re-entrain dust and presumable asbestos.

To date, very little research has been devoted to these areas and concerns. This paper discusses data that are largely observational in nature but are illustrative of general trends of interest to those individuals dealing with the day-to-day problems of asbestos in buildings.

DUST

In the future, surface sampling (dust sampling) of some type may be the most useful indicator of contamination, because of the unpredictable, episodic releases of asbestos from ACM. Currently, there are a variety of techniques being studied for sampling and analysis of dust. The new committee of the National Asbestos Council will evaluate various methods of surface sampling and analysis. The background for most of these techniques relates to intense work done years ago for the nuclear industry.[3] From the nuclear work, which related to the ability of various surface sampling techniques to recover small particles from surfaces, it was observed that extremely small particles (below two micrometers) require local airflow (sampling) velocities near 100 miles per hour to overcome surface attraction forces and dislodge and to recover the particles for sampling. Many asbestos fibers observed by TEM are in this size range. To achieve this high sampling velocity, a special micro-vac sampling cowl, called a SMAIR sampler, was used. This cowl cap utilized specially designed orifices drilled at unique angles to achieve the high velocities. This microvac technique differs from the ones used today for asbestos sampling and is undoubtedly more efficient although less practical. Additional work on the current microvac technique is needed to fully characterize its performance on a variety of surfaces and under a range of loadings.

In Figure 1, the efficiencies of three sampling techniques studied for nuclear applications are demonstrated for three particle size fractions. It is logical that these relationships should be generally similar to those for asbestos, although no data are available to support or refute this. Again, asbestos fibers and structures fall within the size ranges addressed by Figure 1. Most airborne asbestos structures in post-abatement monitoring studies performed by EPA are shorter than five micrometers; in fact, 99 percent fall below five micrometers using the 3:1 minimum aspect ratio PCM counting rules. This relates to the EPA preference not to recommend PCM for post-abatement asbestos clearance use and instead recommends TEM. We assume that dust-laden samples contain generally similar size distributions, except that the smaller individual fibers may have agglomerated into matrices (a mixture of fibers and other materials, stuck together); however, dust particles may not represent airborne particles as the dust size distribution may reflect larger aggregates that have preferentially settled. Additional data evaluation is needed on this topic.

In Figure 2, typical asbestos values in the surface dust are presented for a variety of sites containing ACM. As expected, a wide range is evident, from 6000 asbestos structures/ft^2 to over four billion/ft^2, averaging near one million asbestos structures/ft^2. In addition to areas that are normally cleaned on a routine basis, asbestos dust accumulates in normally-unaccessible areas, which may pose a significant health threat to the maintenance worker. For example, samples taken on the top of ceiling tiles in an ACM area were found to contain as high as 1.2 billion asbestos structures/ft^2. If a worker

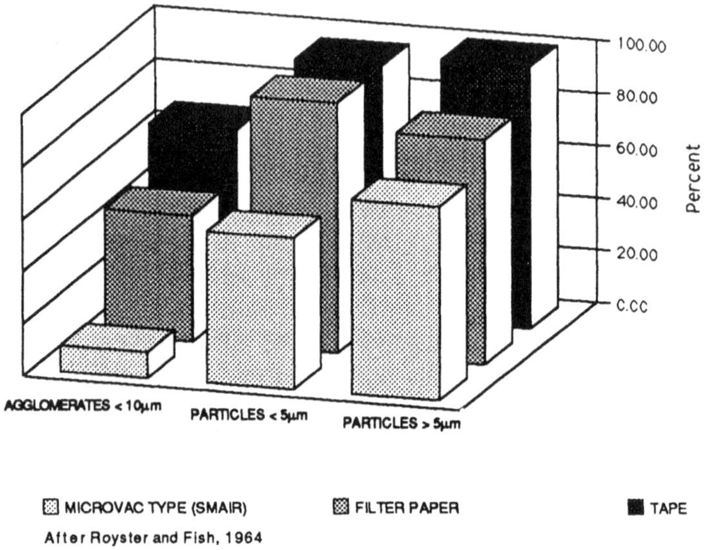

Figure 1. Efficiency of dust collection on hard surfaces.

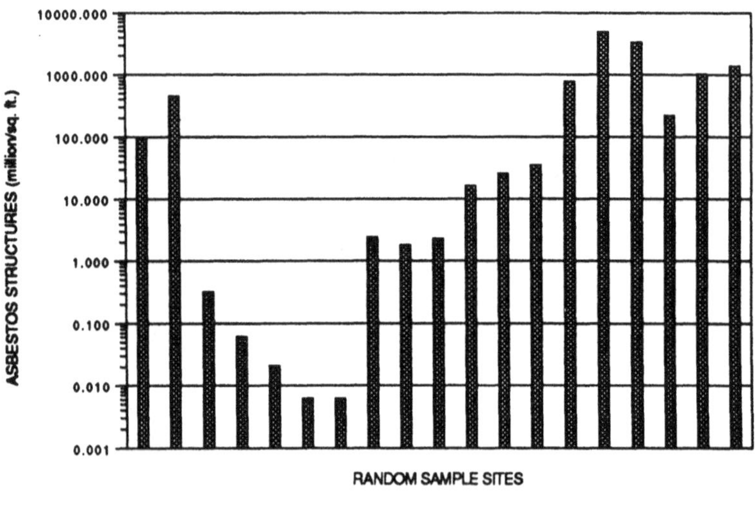

Figure 2. Dust analyses typical values.

unknowingly pops out one of those tiles, serious contamination and exposure could result. Similarly, asbestos-laden-dust in inaccessible ductwork may present a comparable scenario.

CARPET

Carpet acts as a reservoir for asbestos. Two major types of sampling techniques have been attempted on carpet:

1) microvac—use of a) an open cassette or b) a closed cassette with a short piece of Tygon tubing as a sampling nozzle, and
2) bulk carpet sonification—removal of a plug of carpet and ultra-sonification.

The sonification technique sacrifices a small bulk sample of the actual carpet for destructive analysis. Although the sonification technique should provide a better estimate of contamination, it is obviously not going to become popular with building owners, who are unlikely to appreciate the utility of sacrificing pieces of their carpets for testing.

Typical values for microvac are shown in Figure 3 and Figure 4 illustrates sonification values. Not unlike the dust data, the carpet microvac values range from about 8000 asbestos structures/ft^2 to two billion structures/ft^2, averaging about one million asbestos structures/ft^2. Bulk sample carpet sonification (Figure 4) produced values starting at 30 million/ft^2 and ranging to four billion/ft^2, averaging about 100 million/ft^2. Figure 5 shows carpet data averaged for each of 12 sites. In two of the 12, both sonification and microvac analyses were made. The sonification values were 100 times and 20 times higher than the respective microvac analyses.

In an ACM-containing office building, samples were taken in a number of similar rooms in which the owner was able to document the last date that the cleaning service had cleaned that particular area. Assuming somewhat uniform conditions existed from one room to the next, the data in Figures 6 to 8 are useful only to the extent of illustrating the general trend of asbestos accumulation with time, both in the dust and in the carpet.

TYPICAL INDOOR AIR ASBESTOS LEVELS

In Figure 9, typical air levels are presented for some of the sites where dust and carpet samples were obtained. On the day the samples were taken,

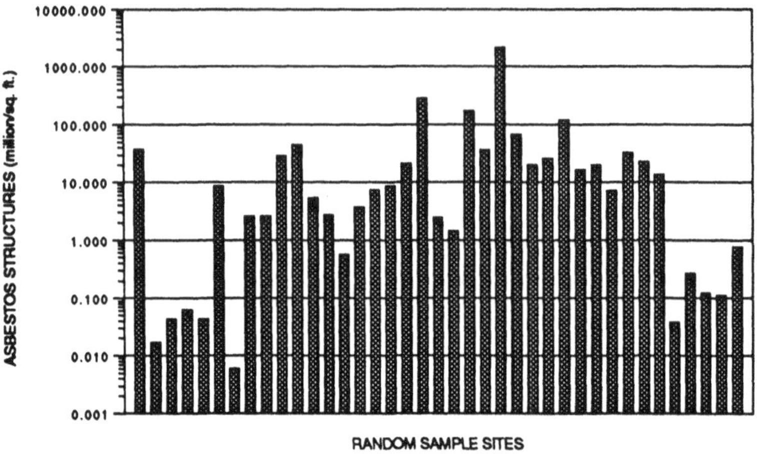

Figure 3. Carpet microvac analyses typical values.

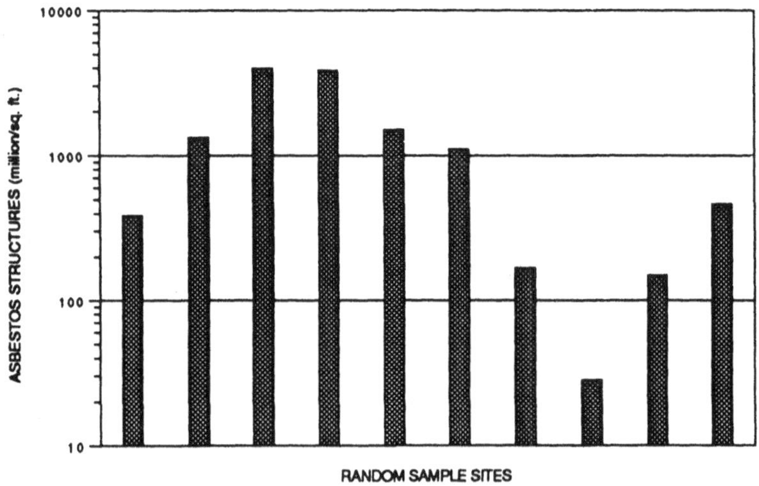

Figure 4. Carpet sonification analyses typical levels.

most of the values were quite low, but three of the sites had elevated levels. As a reference, typical outdoor (ambient) asbestos levels are near 0.005 asbestos structures/cc.

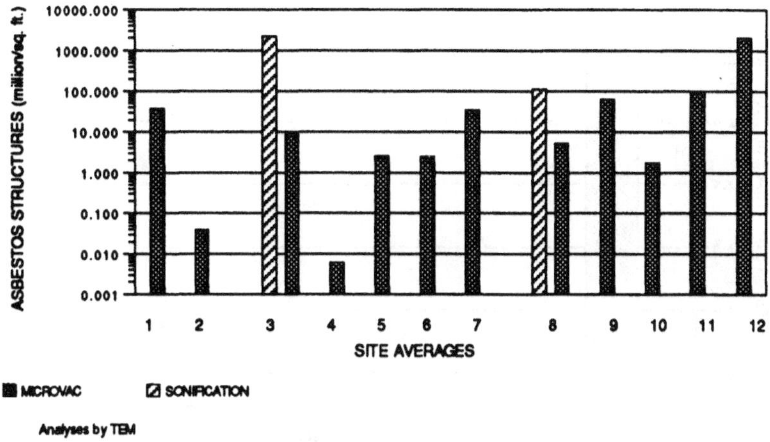

Figure 5. Comparison of carpet microvac versus sonification.

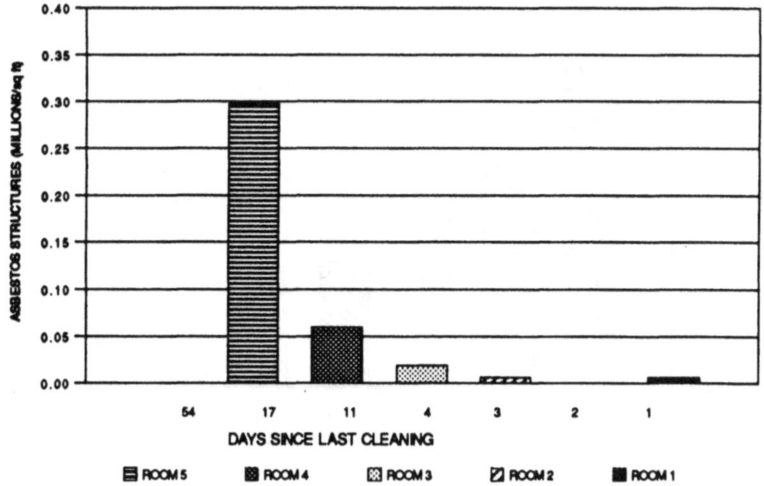

Figure 6. Effect of time since cleaning surface dust.

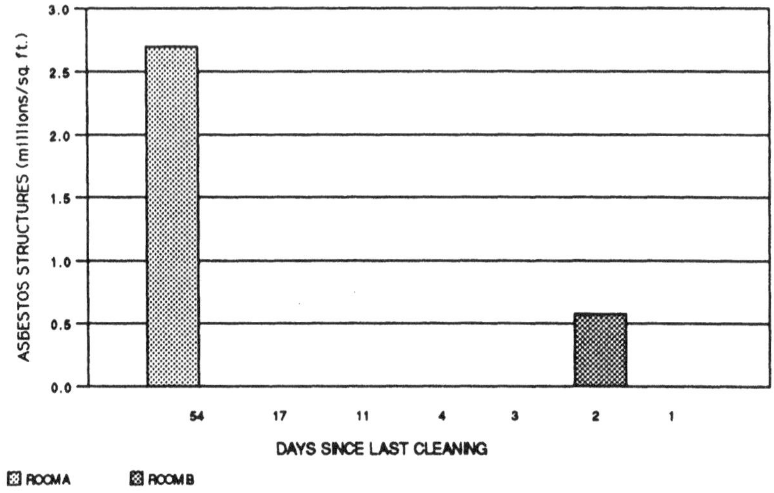

Figure 7. Effect of time since cleaning carpet.

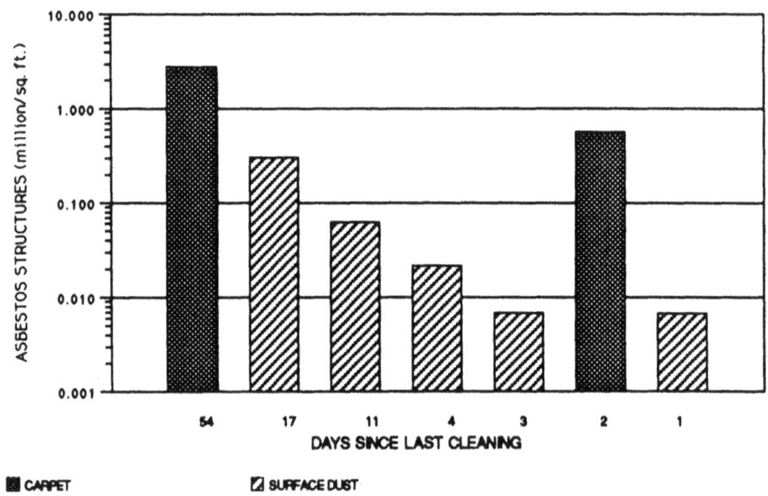

Figure 8. Effect of time since cleaning.

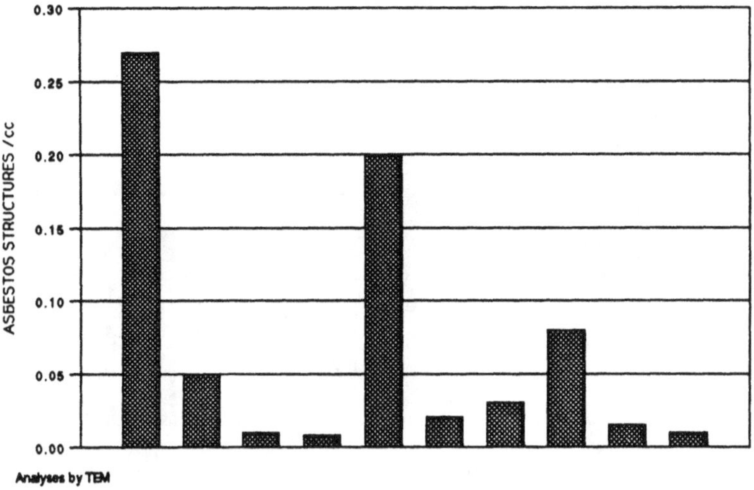

Figure 9. Typical air levels, passive sampling.

ASBESTOS VACUUMS

Most of the published information on the effectiveness of vacuum cleaners for use on asbestos was done at Texas A&M University by Wilson,[4] Daniels,[5] Johnston and Clapp,[6] and Johnston.[7] In the series of studies done by these researchers for NIOSH, they tested the effectiveness of nine vacuums (seven different manufacturers), which were designed for asbestos use. All of the systems used multiple filters. Five of the nine were classified as high efficiency (HEPA). Johnston and Clapp placed the vacuums, one by one, in a clean room. The intakes were attached through a sealed connector to an inlet hopper in the adjacent room. The vacuum motor was started and measured amounts of "pure" chrysotile asbestos were fed into the intake hopper and drawn into the vacuum cleaner located in the adjacent "clean" room. Samples were collected at various locations inside the "clean" room while the vacuum was operating. Analyses were made by PCM and reported as asbestos fibers, which was a valid presentation since only "pure" asbestos was introduced into the vacuum inlet.

The results are shown graphically in Figure 10. Significantly, only two of the nine vacuums tested met the now-current OSHA action level of 0.2 fibers/cc by PCM. In reality, the performance of eight of the nine systems was disappointing and would result in unwarranted exposures during normal

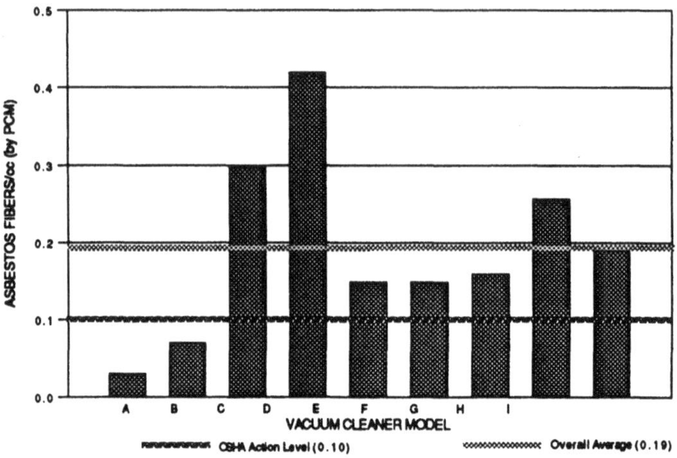

Figure 10. Asbestos vacuum cleaner emissions—molded comparison.

use. The fact that the results were alarming using PCM underestimates the exposure that would have been seen if TEM had been used as the analytical technique. Johnston's tests were conducted around 1980; hopefully, significant improvements in design and performance have been incorporated by manufacturers in the past eight years.

CARPET VACUUMING DATA

Research is just now beginning on evaluation of exposures during actual carpet vacuuming operations Karaffa, et al.,[8–9] in an EPA study, used both PCM and TEM to compare conventional and HEPA vacuum performance at two sites, which we later learned had relatively lower levels of asbestos in the carpet than in other studies. The TEM data, not shown here, suffered from the small air volumes sampled during the brief time it took for the custodial worker to vacuum the area; few asbestos structures were seen on the filters, therefore, most asbestos counts by TEM were below the limit of detection. The study is currently being redesigned to improve TEM-sampling sensitivity. From the low TEM counts, it is clear that no gross contamination was occurring as a result of use of either the conventional or the HEPA vacuum at these two ACM sites. The PCM data (Figure 11) were

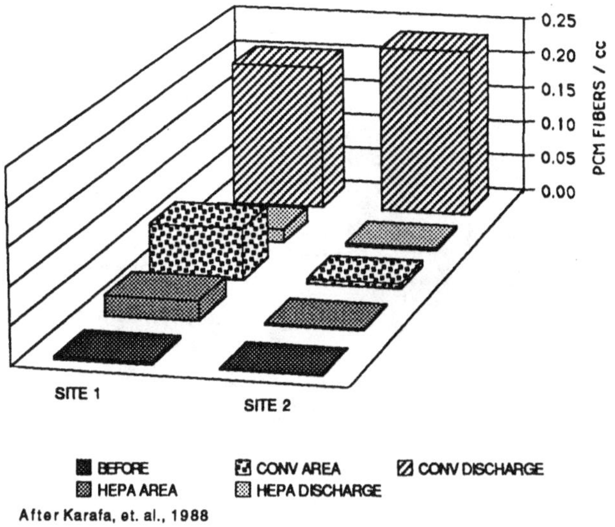

Figure 11. Carpet vacuuming PCM averages.

more definitive and illustrative of the relative performance between the two types of vacuum systems. The observable difference in fiber release (not necessarily asbestos) between the two systems was statistically significant.

McCrone Environmental Services conducted a small, observational study on carpet vacuuming.[10] Although the study was not statistically designed, the resulting data trends are interesting and support the logic increased asbestos levels both at the vacuum discharge and at the breathing zone of the custodial worker (Figure 12). The HEPA was more successful in containing asbestos emissions, although it is interesting that the breathing zone levels were slightly higher than the vacuum discharge levels, possibly indicating that the mechanical action of the vacuum, plus the activity within the room, released asbestos from the carpet. In Figure 13, the asbestos levels in carpets are compared in 1) two adjacent offices and 2) before and after HEPA vacuuming in a third office. Although these values are based upon single data points, the apparent trends are interesting and deserving of further study.

FUTURE RESEARCH

The Environmental Protection Agency Office of Research and Development, in concert with EPA Office of Toxic Substances, NIOSH, and support

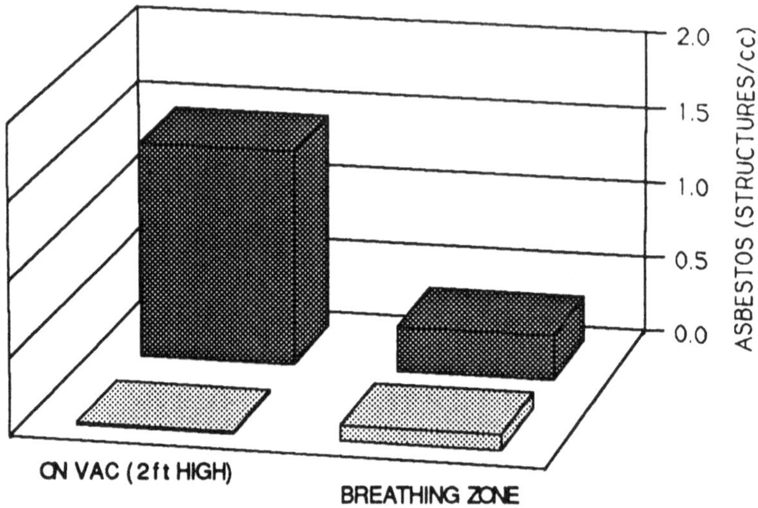

Figure 12. Carpet vacuuming air levels.

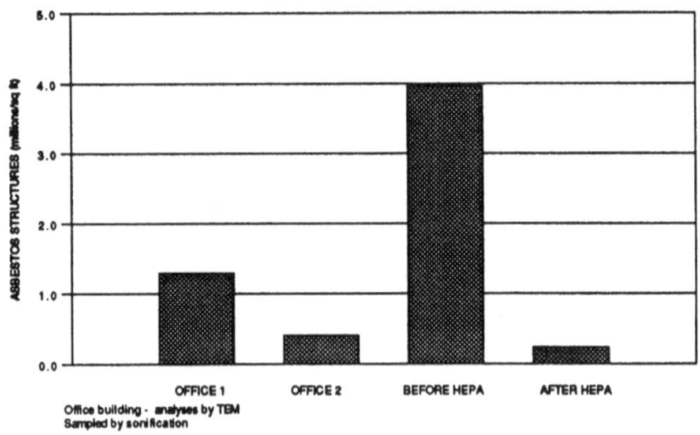

Figure 13. Carpet vacuuming.

contractors, is dedicated to improving the state of the art concerning guidance to the public on effective technologies for Operation and Maintenance activities in buildings containing asbestos. The results of this research will not only be useful to schools, but also to all ACM buildings, including residences. The O&M-related research is a significant component of current EPA research on engineering controls for asbestos. This year, EPA intends to study the O&M activities that we consider to have the greatest potential for worker exposure to asbestos (carpet vacuuming and dusting/cleaning are high on that list). The goal of this research, to be conducted at a captive site where we can best control the many variables, is to evaluate the adequacy of existing control system and, if necessary, to modify or develop technologies to assure control system performance.

REFERENCES

1. **U.S. Environmental Protection Agency,** *Airborne Asbestos Health Assessment Update,* Report EPA/600/8-84/003F, Environmental Criteria and Assessment Office, Research Triangle Park, NC. June, 1986.
2. **Bell, John J.,** *Environmental Engineering News,* School of Civil Engineering, Purdue University, Lafayette, IN. January, 1988.
3. **Royster, George W. and Birney R. Fish,** *Techniques for Assessing "Removable" Surface Contamination,* Proceedings of a symposium on Surface Contamination held at Gatlinburg, Tennessee in June, 1964, Edited by B.R. Fish, Pergamon Press Ltd., New York, NY. 1967.
4. **Wilson, J.T., Jr.,** *Evaluation of Asbestos Fiber Content in Discharge Air from a Portable Canister Vacuum Cleaner,* Texas A&M University, College Station, TX. May, 1975.
5. **Daniels, S.H.,** *Evaluation of Asbestos Dust in Discharge Air from a High-Velocity, Low-Volume Vacuum System,* MS Thesis, Texas A&M University, College Station, TX. December, 1977.
6. **Johnston, W.L. and D.E. Clapp,** "An Evaluation of Vacuum Equipment for Collection of Asbestos Waste," NIOSH Technical Report, DHHS No. 80-137, Cincinnati, OH. September, 1980.
7. **Johnston, W.L.,** Evaluation of Asbestos Fibers in the Exhaust Air of Industrial Vacuum Cleaners, *American Association of Industrial Hygiene Journal* **2,** 2, 71, 1985.
8. **Karaffa, Mark A., C.E. Miller, R.A. Amick, J. Chesson, and C. Jackson,** *Preliminary Evaluation of Potential Asbestos Exposure During Carpet Vacuuming in Buildings Containing Friable Asbestos Material,* U.S. Environmental Protection Agency, Unpublished Draft Report, Cincinnati, OH. November, 1987.
9. **Kominsky, J.R., R.W. Freyberg, J. Chesson, W.C. Cain, T.J. Powers, and R.C. Wilmoth.** Evaluation of Two Cleaning Methods for the Removal of Asbestos Fibers from Carpet. *Am. Ind. Hyg. Assoc. J.* **51,** 9, 500–504, 1990.

10. **Millette, J.R.,** *Dust Analysis in Buildings*. Paper presented at the National Asbestos Council Conference, Atlanta, GA. February 19, 1988.

NOTES

* Operations and Management

AUTHOR AFFILIATIONS

** U.S. Environmental Protection Agency, 26 W. Martin Luther King Drive, Cincinnati, OH 45268

*** Vander Wood & Associates, Inc., 5500 Oakbrook Parkway #200, Norcross, GA 30093

Source: Microscope (1991) 39. Used with permission of McCrone Research Institute, Chicago, IL.

APPENDIX 8

Re-entrainment of Asbestos from Dust in a Building with Acoustical Plaster

Dale L. Keyes, Jean Chesson, Steve M. Hays,
Richard L. Hatfield, William M. Ewing,
William E. Longo, and James R. Millette

A recent study of cleaning and athletic activities sheds light on the relationship between asbestos in surface dust and exposure to airborne asbestos. Specifically, a simulation was conducted of (a) playing basketball and tumbling and (b) dry sweeping and dusting at least partially derived from asbestos-containing acoustical plaster. Significant increases in airborne asbestos concentrations were observed during the simulations.

Re-entrainment of dust from floor and other horizontal surfaces is believed to be an important source of airborne asbestos in some buildings.[1–6] This is particularly true for buildings where deterioration of or damage to in-place asbestos-containing material (ACM) has produced significant accumulation of dust laden with asbestos fibers.

This paper details an investigation of asbestos dust re-entrainment in the Greenville, South Carolina YWCA. Asbestos-containing acoustical plaster had been applied to ceilings during construction in 1964. The plaster had suffered from structural movement, water damage from roof leaks and physical damage from vandalism. One room (the Gymnasium) had been closed to use for about one year and access to another (the Jolly Room) had been limited for several months due to substantial accumulation of dust and small pieces of debris similar in appearance to the plaster. The rooms were scheduled for ACM removal; this study was conducted just prior to abatement. Athletic activities in the Gym and cleaning activities in the Jolly Room (used as a storage area) are typical of activities that might cause asbestos fibers to become airborne. These activities were investigated as possible causes of re-entrainment in each of the rooms.

HYPOTHESIS

If re-entrainment is to occur, sufficient energy must be imparted to the surface dust for fibers to become airborne. Moreover, for the concentrations of airborne asbestos to rise significantly, the amount of asbestos in dust must be substantial. It was postulated that the amount of asbestos in the Gym and Jolly Room was substantial as reflected by the visible appearance of the dust, and that the bouncing of balls and other activities in the Gym and the sweeping and dusting of the Jolly Room, would re-entrain the dust and elevate airborne asbestos in each room.

RESEARCH DESIGN

The approach to investigating these hypotheses was to conduct controlled simulations of (a) athletic activities in the Gym and (b) cleaning activities in the Jolly Room.

The situation was investigated, were occupants and workers are unaware that asbestos is present in the dust and thus take no special precautions to minimize dust suspension. Only settled dust was disturbed; no in-place ACM was contacted.

Air samples were collected before, during and after the simulated activities and analyzed for asbestos using transmission electron microscopy (TEM). The "after" phase started three hours following the activity and was designed to detect the effect of fiber settling on airborne concentrations. Elevations in airborne asbestos concentrations during the activities would indicate that asbestos was re-entrained; lower concentrations during the "after" phase would suggest the rate of fiber settling.

The study was designed to control false positive and false negative error rates. A false positive error occurs when one concludes that airborne asbestos levels have increased when in fact they are unchanged. A false negative error occurs when one concludes that airborne asbestos levels have not increased when in fact they have. The false positive error rate was maintained at five percent by using a statistical significance level of 0.05. Sufficient samples were collected so that the false negative error rate would be less than five percent for increases of 10-fold or greater. In other words, the study is likely to conclude that there is a significant difference between two airborne asbestos concentrations if one is 10 times greater than the other. Smaller increases, however, may not be detected.

Dust samples were also collected from the various horizontal surfaces in each room. Each type of surface was sampled separately using a random scheme. The samples were then analyzed by TEM for asbestos. A weighted average (by surface area) was calculated to provide a composite index of "asbestos dustiness" in units of asbestos structures per square centimeter.

Figures 1 and 2 depict the configuration of each room. The Gym is about 3,850 square feet in size and encompasses roughly 96,000 cubic feet; the Jolly Room is much smaller–about 1,800 square feet and 22,000 cubic feet. The figures show the types of floor coverings, furnishings and other materials found in the rooms.

METHODS

Site Preparation. Each room was sealed from other parts of the building by locking doors and constructing critical barriers of polyethylene plastic at doorways and windows. In addition, the forced-air HVAC system in the Gym had previously been turned off and the hot water radiator system in the Jolly Room was deactivated. A wet decontamination unit with a fixed, sealed door was constructed at an entry to each room. Although two HEPA air filtration units were operated in the recirculation mode at the conclusion of each experiment, that is, following the "after" period, the abatement contractor was requested to treat the rooms as potentially contaminated sites while preparing them for subsequent abatement.

Activity Simulation. Activities in the Gym included two-on-two basketball games, tumbling on mats and the use of gymnastic equipment (parallel bars, pommel horse and uneven bars). These activities lasted roughly two hours. Activities in the Jolly Room lasted approximately one hour and consisted of dry dusting (with cloths) furniture, stored items and window sills; dry-sweeping floors with brooms; and stacking the cleaned furnishings and equipment.

Figures 3 and 4 are photographs depicting these activities. As shown, participants in the simulations wore protective clothing and powered air-purifying respirators.

Dust Sampling. Dust samples were collected from various surfaces (e.g., wood floor, carpet, fabric upholstery, foam rubber, canvas cot, plastic mat, metal radiator cover) using the microvacuum technique.[6] Square areas of surface (100 cm^2) were vacuumed using a template and a low volume pump with 25 mm sampling cassettes containing MCE (0.45 μm pore size) filters.

Air Sampling. Area (stationary) as well as personal (on workers) samples were collected. High-volume pumps were used for area samples; low-volume pumps were used for personal samples. The location of the area samplers is shown in Figures 1 (Gymnasium) and 2 (Jolly Room). Collection media consisted of mixed cellulose ester (MCE) filters with 0.45 μm pore

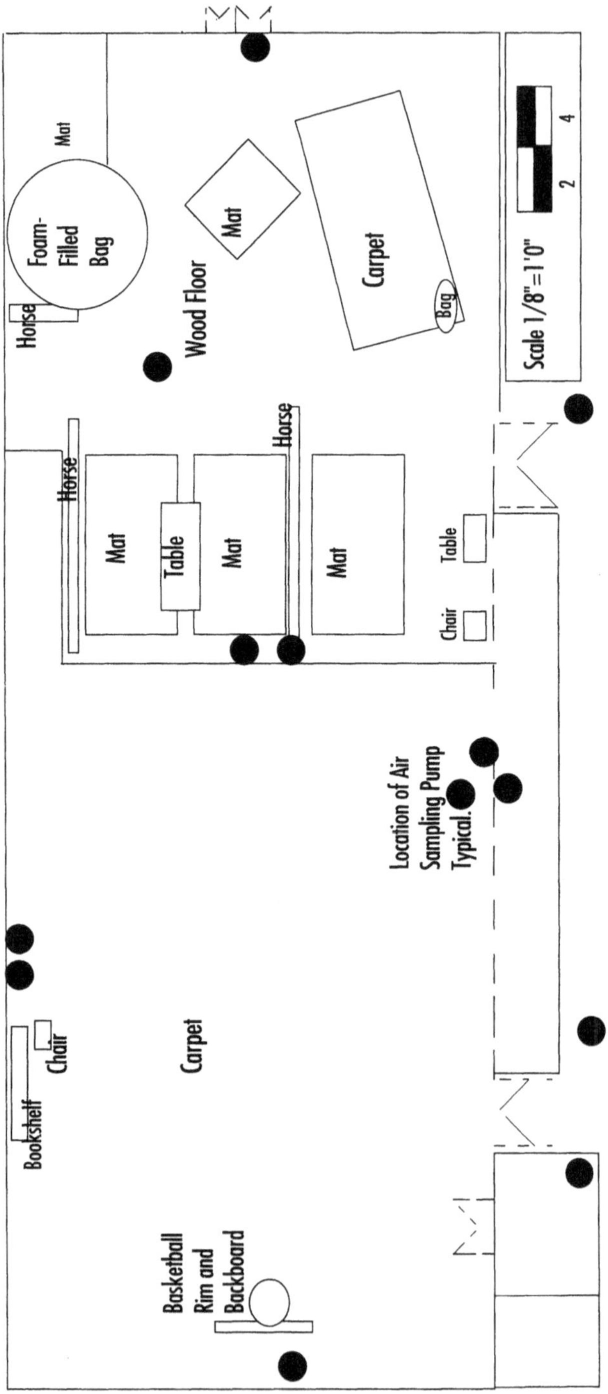

Figure 1. The Gymnasium.

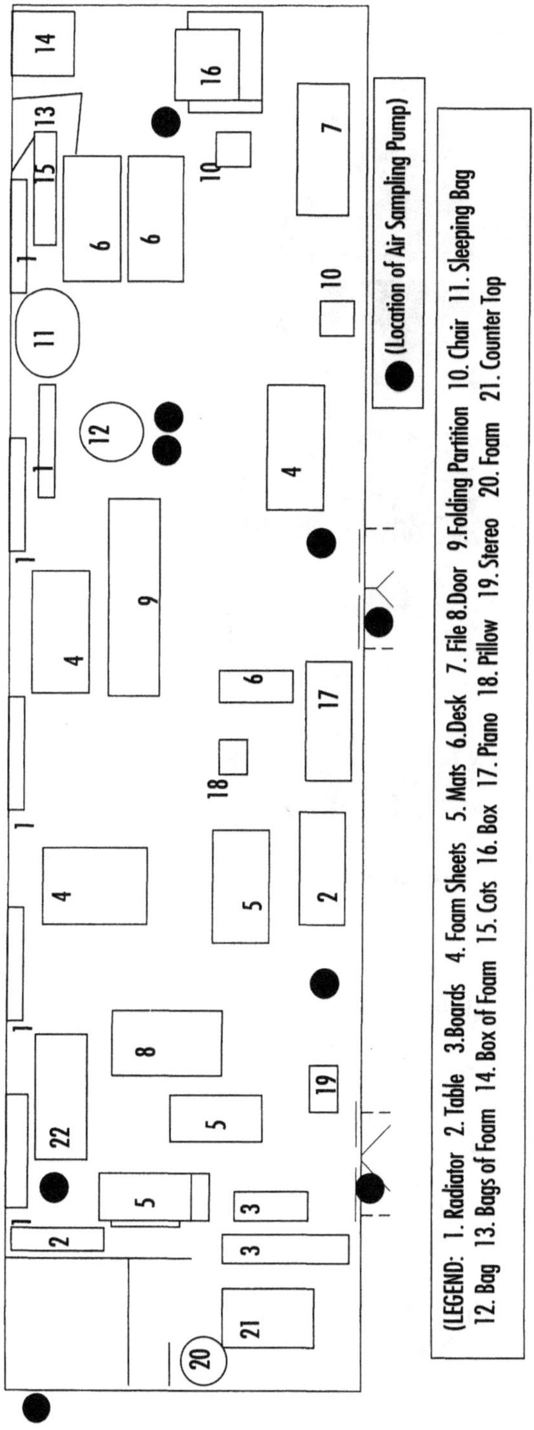

Figure 2. The Jolly Room.

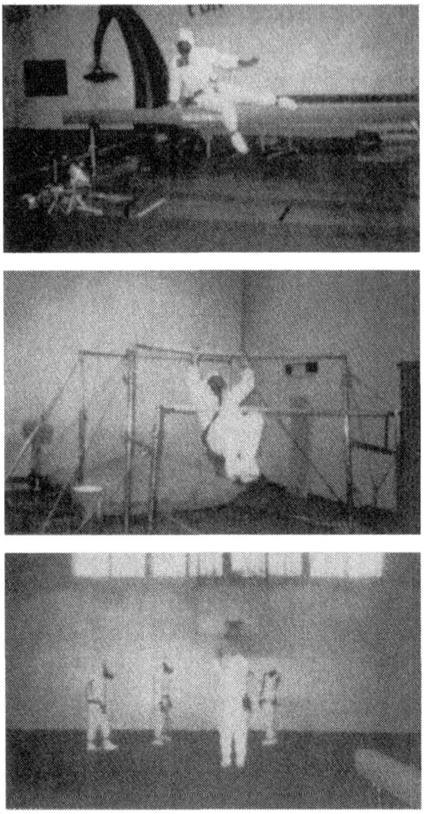

Figure 3. Simulated activities in the Gymnasium.

size. Flow rates were generally eight to 10 liters per minute (lpm) for the high flow pumps and two to three lpm for the low-volume pumps. Flow rates were measured before and after sampling.

Sampling Analysis. All air and dust samples were analyzed using TEM methods. An indirect sample preparation (a slightly modified Level II analysis based on Yamate, 1984[7]) was used for the air samples and all resolvable asbestos structures were counted. The analysis of dust samples followed the draft U.S. Environmental Protection Agency (EPA) method.[6] Method Analytical Services, Inc. analyzed the samples.

Quality Assurance. The quality assurance program consisted of blank samples (both field and laboratory blanks), replicate analyses of randomly

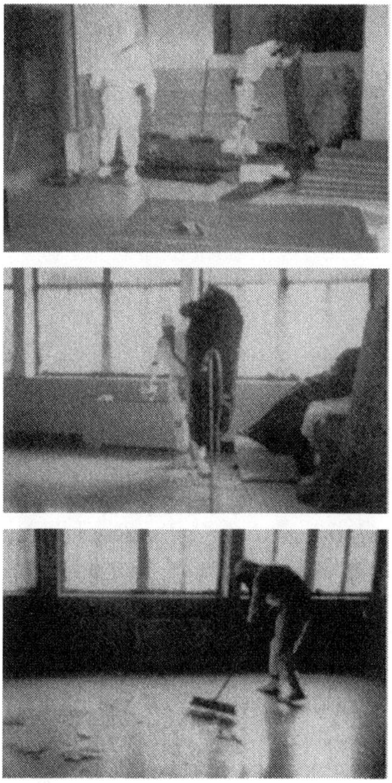

Figure 4. Simulated activities in the Jolly Room.

selected samples by a second laboratory (McCrone Environmental Services, Inc.), and extensive documentation of sampling and analysis conditions, including the use of chain-of-custody procedures.

Statistical Analysis. Differences in airborne asbestos levels between phases were tested with a standard one-way analysis of variance using the transformation $\log_{10}$ (x + 0.0025) where x is the data value. A small constant (0.0025 used here) was needed because there were some zero observations and the log of zero is mathematically undefined. The logarithmic transformation is a standard procedure for air quality data which commonly approximate a log normal distribution. The transformation tends to equalize variances and permit the use of standard statistical tests that would otherwise be inappropriate. The Scheffé test, a standard statistical test, was used to determine the significance of particular pair-wise comparisons. "Geometric" means were calculated as 10^y-0.0025, where y is the arithmetic mean of the log-transformed data. This mean, which differs slightly from the usual definition of the geometric mean because of the addition of a constant in the transformation (and its subsequent substraction), is the appropriate quantity to report because it is the quantity tested by the analysis of variance. Using the analysis of variance to test for the difference between geometric means is equivalent to testing whether the ratio of the arithmetic means differs from one.

RESULTS

Tables 1 and 2 show the dust analysis results for the Gymnasium and Jolly Room, respectively. The average amount of asbestos on surfaces in the Jolly Room was considerably higher than the amount on surfaces in the Gym (870,000 vs 9,700 structures per square centimeter).

Tables 3 (Gymnasium) and 4 (Jolly Room) on the following pages summarize the study results in terms of average concentrations of airborne asbestos in each of the three phases of the experiments. Shown are the arithmetic means, arithmetic standard deviations and the geometric means. Tables 3 and 4 also report the results of testing differences in means among the three experiment phases for statistical significance.

As shown in Table 3, the athletic activities produced modestly elevated concentrations of airborne asbestos as measured by the area samplers (0.098 vs 0.0 s/cm^3) and substantially higher levels as measured by the personal samplers (0.484 s/cm^3). The elevations (area samplers) were statistically significant as judged by the conventional measure of significance (i.e., a p value of 0.05 or lower). Concentrations decreased after the three-hour period presumably due to fiber settling and the "after" level was not significantly different from the "before" level.

Table 4 reveals that cleaning produced higher elevated concentrations of airborne asbestos than did the athletic activities as measured by both the area samplers (2.95 s/cm^3) and personal samplers (2.74 s/cm^3). Again, the increase as measured by the area samplers was statistically significant when compared with the "before" level. The three-hour settling period resulted in a lower average concentration (0.341 s/cm^3), which, again, was not statistically different from the "before" level.

Although the study was not designed to compare asbestos concentrations between experiments, the concentrations during cleaning activities are higher by at least an order of magnitude than those during athletic activities (area samples).

Table 1. Summary of dust results for the Gym.

SURFACE TYPE & SAMPLE	ASBESTOS DENSITY (s/cm^2)
Carpet and Foam Bag (61% of Total Surface)	
300	1.04×10^4
311	6.26×10^3
314	2.43×10^3
Arithmetic Mean	6.36×10^3
Mat (11% of Total Surface)	
302	0
307	1.51×10^4
309	3.46×10^3
Arithmetic Mean	6.19×10^3
Wood Floor (28% of Total Surface)	
301	3.27×10^4
305	3.04×10^3
306	1.90×10^4
Arithmetic Mean	1.82×10^4
Weighted Mean	9.7×10^3 s/cm^2

With respect to the quality assurance aspects of the study, no asbestos was found on any of the field or laboratory blanks analyzed. In addition, replicate analysis by the quality assurance of laboratory of six samples selected in a stratified, random manner produced excellent agreement between the two laboratories: a regression line with a slope of 1.03 (1.0 would be perfect agreement) and a correlation coefficient of 0.98 ($p=0.001$).

DISCUSSION

Both athletic activities and cleaning are capable of re-entraining asbestos from surface dust and significantly increasing concentrations of airborne asbestos. The fact that basketball and gymnastics activities produced smaller increases in airborne asbestos than did dry sweeping and dusting may reflect differences in the amount of asbestos in surface dust available for re-entrainment, differences in the volume of air into which the asbestos was entrained

Table 2. Summary of dust results for the Jolly Room.

SURFACE TYPE & SAMPLE #	ASBESTOS DENSITY (s/cm^2)
Hard Furniture (13% of Total Surface)	
317	0
329	5.56×10^2
Arithmetic Mean	2.78×10^2
Soft Furniture (9% of Total Surface)	
320	3.81×10^3
328	2.42×10^4
Arithmetic Mean	1.40×10^4
Sills & Radiators (5% of Total Surface)	
321	0
305	3.44×10^6
Arithmetic Mean	1.72×10^6
Vinyl Floor[a] (72% of Total Surface)	
324	0
330	2.16×10^6
Arithmetic Mean	1.08×10^6
Weighted Mean	8.7×10^5 s/cm^2

a The floor covering was reported to be asbestos-free.

and differences in the effectiveness of re-entrainment processes. Recall that the density of asbestos on surfaces in the Jolly Room was almost 100 times higher than the density in the Gym, and the volume of the Jolly Room is less than one-fourth the volume of the Gym.

It is difficult to interpret the results with respect to health risks. The only health standards currently available are the OSHA standards for asbestos workers[8] and are stated in terms of fibers longer than 5 µm as observed with phase contrast microscopy (PCM). These results are reported in terms of all asbestos structures (i.e., free fibers, fiber bundles and clusters and matrices of asbestos and non-asbestos materials) as observed by TEM. Although a direct comparison between PCM and TEM analytical results is not valid,[9] the results in this study include many asbestos fibers not observable by PCM (about 95 percent of the fibers reported in this study were shorter than 5 µm and/or thinner than 0.25 µm).

Table 3. Statistical analysis of results for the Gym—descriptive statistics (in s/cc).

GROUP	ARITHMETIC MEAN	ARITHMETIC ST. DEV.	"GEOMETRIC" MEAN[a]	NUMBER OF OBSERVATIONS
Before (area)	0.0	0.0	0.0	5
Athletic Activity (area)	0.230	0.264	0.098	6
Athletic Activity (personal)	0.582	0.450	0.484	3
After (area)	0.021	0.031	0.006	5

Scheffé test for differences between geometric means

TEST	STATISTICAL SIGNIFICANCE[b] PROBABILITY
Before vs Athletic Activity (area)	0.005
Before vs After (area)	0.472
After vs Athletic Activity (area)	0.056

a Defined as $10^y - 0.0025$ where y is the arithmetic mean of the transformed data. The transformation was $\log_{10}(x + 0.0025)$ where x is the observed value.

b A small probability indicates that the difference in the two means is unlikely to have occurred by chance.

Note: Analytical sensitivity was approximately 0.03 - 0.06 s/cm^3 for before and after samples and about 0.06 s/cm^3 for the activity samples (area samples) and 0.29 - 1.14 s/cm^3 (personal samples).

Table 4. Statistical analysis of results for the Jolly Room—descriptive statistics (in s/cc).

GROUP	ARITHMETIC MEAN	ARITHMETIC ST. DEV.	"GEOMETRIC" MEAN[a]	NUMBER OF OBSERVATIONS
Before (area)	0.314	0.680	0.013	5
Cleaning (area)	3.123	1.265	2.952	5
Cleaning (personal)	2.781	0.708	2.735	2
After (area)	0.384	0.233	0.341	4

Scheffé test for differences between geometric means

TEST	STATISTICAL SIGNIFICANCE[b] (PROBABILITY)
Before vs Cleaning (area)	0.002
Before vs After	0.068
After vs Cleaning (area)	0.228

a Defined as $10^y - 0.0025$ where y is the arithmetic mean of the transformed data. The transformation was $\log_{10}(x + 0.0025)$ where x is the observed value.

b A small probability indicates that the difference in the two means is unlikely to have occurred by chance.

Note: Analytical sensitivity was approximately 0.03 - 0.04 s/cm^3 for before and after samples and about 0.12 s/cm^3 for the activity samples (area samples) and 0.55 - 1.07 s/cm^3 (personal samples).

The health significance of these relatively small asbestos fibers remains a hotly debated topic.[9–10] Nevertheless, it is believed that avoiding or minimizing the re-entrainment of asbestos from surface dust is prudent. This means that surfaces below in-place ACM should be cleaned regularly to avoid accumulation of asbestos-laden dust and, at least initially, the cleaning should be conducted by trained workers wearing respiratory protection and in a manner designed to minimize re-entrainment. In other words, a comprehensive asbestos operations and maintenance program should be implemented and followed.

ACKNOWLEDGMENTS

The work reported in this article was supported in part by the Plaintiff's Executive Committee of the National Schools' Class Action for Cost Recovery. The authors exercised full control over design and execution of the study and the interpretation and publication of the study results. The following persons and organizations participated in the field portion of the study: Clay Call and Michael Thompson (Diagnostic Engineering, Inc.), Dale Keyes (Environmental Sciences, Inc.), Steve Burch (Gobbell Hays Partners, Inc.), Fred Gaunce and Rich Sturgeon (McCrone Environmental Services, Inc.). Various persons from Marshall Clarke, Architect; and Ballard Abatement Co. also helped facilitate the field work. Finally, the authors wish to thank the Greenville, SC, YWCA and, in particular, Sally Perry, for their cooperation.

REFERENCES

1. **Sawyer, R.N.,** "Asbestos Exposure in a Yale Building." *Environmental Research.* Vol. 13, 146–169, 1977.
2. **Lumley, K.P.S., P.G. Harries and F.J. O'Kelly,** "Buildings Insulated with Sprayed Asbestos: A Potential Hazard." *Annals of Occupational Hygiene.* Vol. 14, 255–257, 1971.
3. **Litzistorf, G., M.P. Guillemin, P. Buffat and F. Iselin,** "Influence of Human Activity on the Airborne Fiber Level in Paraoccupational Environments." *Journal of the Air Pollution Control Association.* Vol. 35(8):836–837, 1985.
4. **Guillemin, M.P., P. Madelaine, G. Litzistorf, P. Buffat, F. Iselin,** "Asbestos in Buildings: The Difficulties of a Reliable Assessment." *Aerosol Science and Technology.* Vol. 11, 221–243, 1989.

5. **Carter, R.F.,** *The Measurement of Asbestos Dust Levels in a Workshop Environment.* AWRE Report No. 0 28/70. United Kingdom Atomic Energy Authority, H.M. Stationery Office, 1970.
6. **Millette, J.R., T. Kremer and R.K. Wheeles,** "Settled Dust Analysis Used in Assessment of Buildings Containing Asbestos." *Microscope.* Vol. 38, 215–220, 1990.
7. **Yamate, G., S.C. Agarwal and R.D. Gibbons,** *Methodology for Measurement of Airborne Asbestos by Electron Microscopy (Draft).* U.S. EPA, Office of Research and Development, Research Triangle Park, NC, 1984.
8. 29, Code of Federal Regulations, 1926.
9. **Chesson, J., J.D. Rench, B.D. Schultz and K.L. Milne,** "Interpretation of Airborne Asbestos Measurements." *Risk Analysis.* Vol. 10(3), 437–447, 1990.
10. Health Effects Institute-Asbestos Research, "Asbestos in Public and Commercial Buildings: A Literature Review and Synthesis of Current Knowledge." Cambridge, MA, 1991.

AUTHOR AFFILIATIONS

Dale L. Keyes, Environmental Sciences, Inc., 105 East Speedway Boulevard, Tucson, AZ 85705 (602)792-0097; Jean Chesson, Australian Resource Assessment Commission, Queen Victoria Terrace, Canberra, ACT 2600, Australia; Steve M. Hays, Gobbell Hays Partners, Inc., 217 Fifth Avenue, North, Nashville, TN 37219; William E. Ewing, Diagnostic Engineering, Inc., 350 Franklin Road, Suite 300, Marietta, GA 30067; Richard L. Hatfield, Law Associates, Inc., 114 Town Park Drive, Suite 300, Kennesaw, GA 30144; William E. Longo, Materials Analytical Services, Inc., 3597 Parkway Lane, Suite 250, Norcross, GA 30013; James R. Millette, Millette, Vander Wood and Associates, Inc., 5500 Oak Brook Parkway, Suite 200, Norcross, GA 30093.

Source: Environmental Choices Technical Supplement, July/August 1992. Used with permission of The Environmental Information Association, Atlanta, GA.

APPENDIX 9

Scanning Electron Microscopy of Post-it™ Notes Used for Environmental Sampling

Post-it™ notes are readily available and easily used in the field to collect particles of dust. The notes were used to collect samples for SEM examination in three situations involving asbestos contamination, power plant dust and possible home contamination by a fly ash pile containing lead.

J.R. Millette,* R.S. Brown,*
J. Barnett,** and M.D. Mount***

In the course of environmental audits or building surveys, the investigator may come across areas containing particles of dust and fine debris that need further identification. The scanning electron microscope with energy dispersive X-ray analysis (SEM-EDS) capability is one instrument that can be used to locate and study particles containing lead, cadmium, chromium or matrices containing asbestos fibers. The information gathered may be used to help determine where the dust particles originated.

A number of techniques are available for collecting dust samples. Scraping material into a pile and with thumb and forefinger pinching a bit into a container, wet wiping, microvacuuming with an air sampling cassette, and direct collecting using double stick tape and packing tape are useful. But the semi-sticky adhesive of the Post-it™ note has several advantages. One trait used in the collection of trace evidence and for the isolation of particles, such as in the analysis of sands,[1] is the ability to easily transfer particles from the Post-it™ to another media for analysis. In this way an interesting particle located under the SEM-EDS can be easily extracted for further examination under the polarized light microscope (PLM) and vice versa.

Perhaps the most important advantage of Post-it™ notes are their general availability. Anyone who comes upon curious dusty conditions—hygienist, facility manager or building owner—can readily find Post-it™ notes for sampling. This article describes three instances in which Post-it™ notes were used to sample dusts in the field for subsequent examination by SEM.

SAMPLING PROCEDURES

As described by Fisk,[2] the Post-it™ sampling procedure simple requires pressing the sticky edge of the note on the dust to be sampled, folding the note onto itself so the particles and sticky portion are enclosed, and writing information such as location, date and sampler on the note. In the field, it is appropriate to then slip each note into its own sealed plastic bag. Small plastic bags, 1 inch by 3 inch, are most convenient.

STUDY SITUATIONS

Post-it™ notes were used to investigate three different dust situations. The first was during an assessment of an office building for asbestos contamination. A very light amount of dust was evident on the top of an air induction unit in a room that contained only one asbestos-containing material (ACM). A thin (1 inch) strip of ACM was inside the bottom of the unit. Post-it™ samples were taken from both the surface of the unit and the surface of the ACM line to determine if the dust on the surface of the induction unit could contain asbestos from inside the unit.

The second situation involved settled dust within a power plant that generated steam via coal-fired boilers. Post-it™ samples were collected from various horizontal surfaces within the plant such as light fixtures, desk tops, hand rails, pipe (insulation) runs and duct work. The samples were analyzed for heavy metals such as lead, cadmium or arsenic.

The third situation was a study of dust particles accumulated inside a home to see if they were similar to those in a nearby pile of fly ash, which had been shown to contain lead, cadmium and chromium.

EQUIPMENT AND METHOD

The microscope used was a Philips 525M scanning electron microscope with detectors for backscattered electrons (BE), secondary electrons (SE) and an EDAX PV9900 Microanalysis/Image Processing System with an ECON-4 windowless light element detection unit.

The Post-it™ note was opened and examined under a stereomicroscope for debris. An area containing a moderate amount of particle loading was selected and cut out using a clean single-edged razor blade (the square was about 5mm × 5mm). An Avery "Spot of Glue" adhesive tab was transferred to a carbon planchet. The excised Post-it™ square was then outlined with carbon paint to create a conductive path to the carbon planchet. The planchet was mounted on an aluminum pin mount and carbon coated.

SEM examination was carried out initially at a low magnification, usually 150 diameters. Large particles and clusters of large particles were searched at higher magnification for fibers and particles of interest. The examination was completed by examining the entire Post-it™ square at a magnification of 600 diameters.

Analysis was carried out by observing the sample's BE image and SE image simultaneously on two different monitors. This allowed for rapid identification of heavy elements that occur in fly ash particles.

RESULTS

Figure 1 (secondary electron image) shows the surface of the sticky portion of an unused Post-it™. The globules of glue and fibers of cellulose (paper) are clearly visible. The blank Post-it™ was found to contain high levels of zinc/chlorine particles, which appear as bright areas in the backscatter image—Figure 2. It may be speculated that their presence controls the adhesive properties of the glue. No particles containing asbestos, lead, chromium or cadmium were seen in the blank.

Figure 3 and Figure 4 show the secondary electron images of a matrix of chrysotile and mineral wool found on the surface of an air induction unit, which was consistent with the material collected on a Post-it™ sampling from inside the unit—Figure 5.

Figure 6 shows the secondary electron image of fly ash from settled dust around the coal-fired power plant (7th floor, inside building). Figure 7 shows the X-ray analysis spectrum of a lead particle in the fly ash.

Figure 8 shows the backscatter and secondary electron images of a fly ash particle from the pile of suspect source material. The X-ray spectrum containing lead, iron and manganese is shown in Figure 9. Fly ash particles found in the dust collected on a Post-it™ in the house were less distinctive than the particles from the pile. The "in-house" fly ash appeared to be more

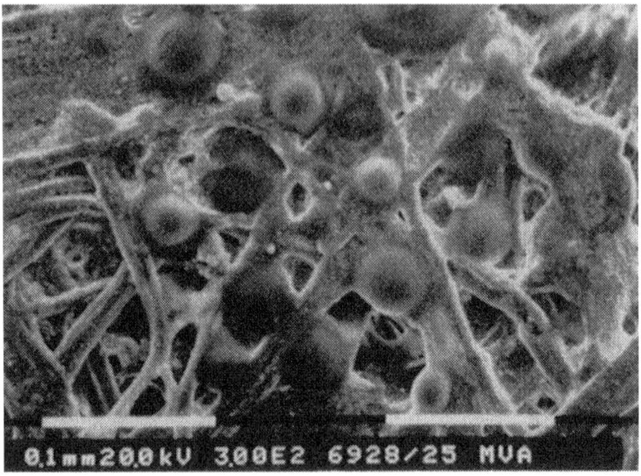

Figure 1. Secondary electron image showing glue side of Post-it™. Bar = 0.1mm.

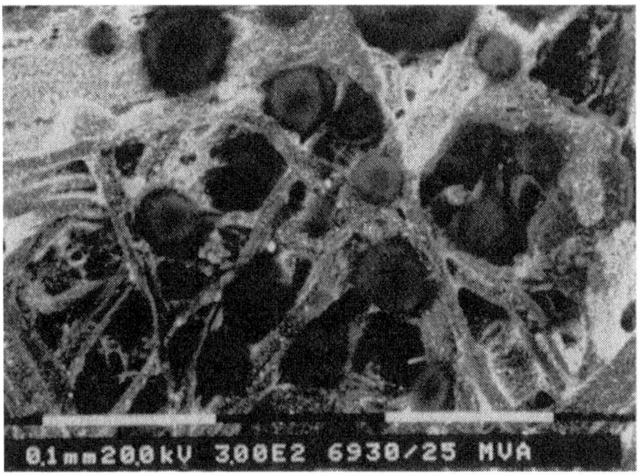

Figure 2. Backscatter electron image showing zinc/chlorine particles (bright areas) on glue side of Post-it™. Bar = 0.1mm.

Figure 3. Secondary electron image showing mineral wool, chrysotile and binder. Post-it™ collected from surface of air induction unit. Bar = 0.1mm.

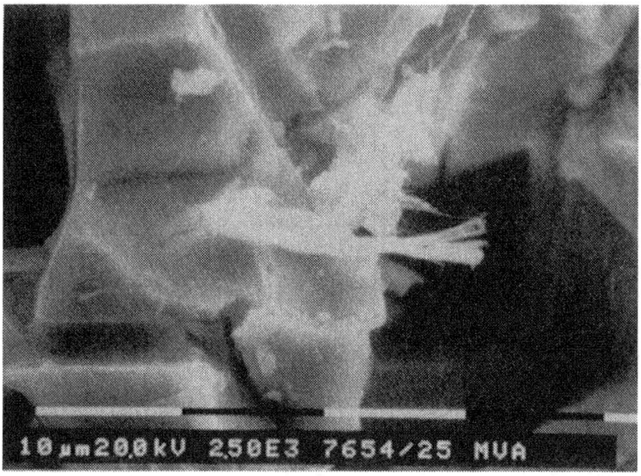

Figure 4. Secondary electron image of chrysotile bundle in Figure 3. Bar = 10µm.

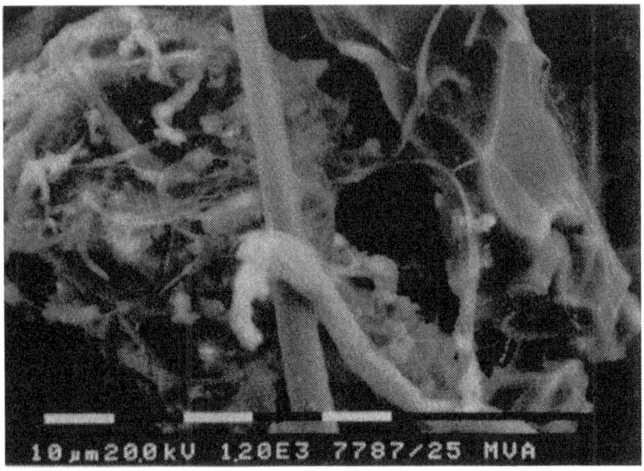

Figure 5. Secondary electron image showing mineral wool, chrysotile and binder. Post-it™ collected from ACM inside the air induction unit. Bar = 10µm.

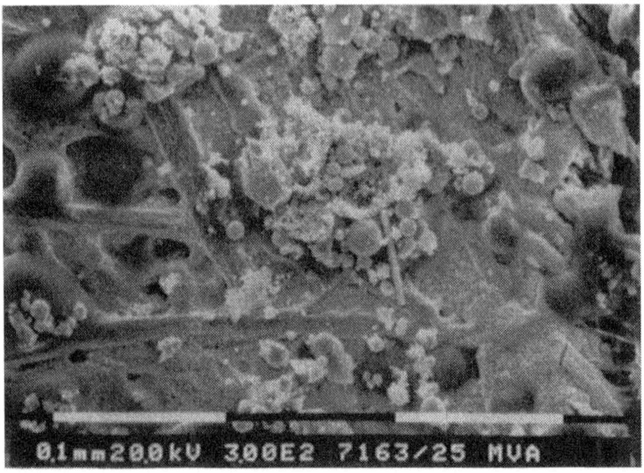

Figure 6. Secondary electron image of fly ash on Post-it™, magnification = 300X. Bar = 0.1mm.

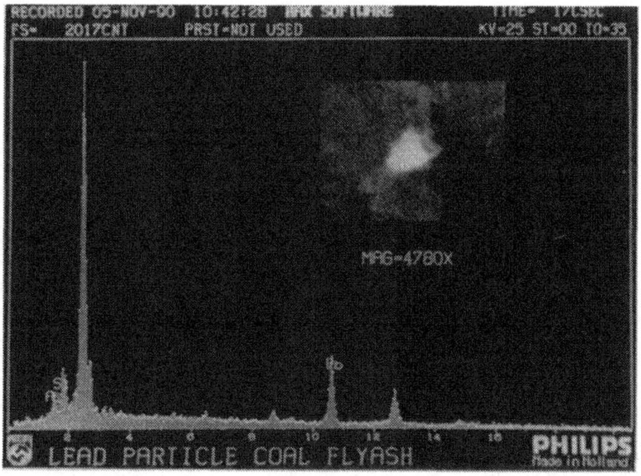

Figure 7. X-ray spectrum of lead particles on Post-it™ of fly ash from coal-powered plant. Inset is micrograph of particle analyzed.

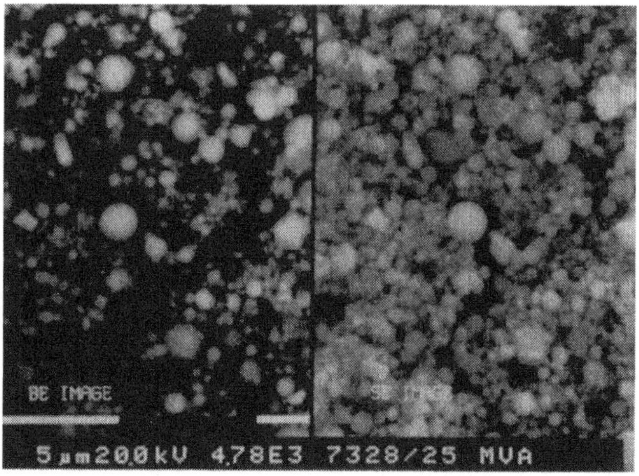

Figure 8. Backscatter electron image (L) and secondary electron image (R) of fly ash particles from pile. Bar = 5µm.

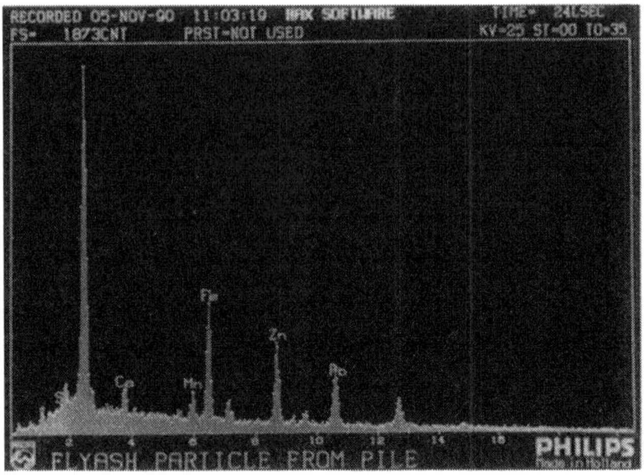

Figure 9. X-ray spectrum of a fly ash particle in Figure 8.

damaged and mixed with other material. However, some particles from the house dust, as shown in Figure 10, were consistent with the fly ash from the pile. The spectrum profiles, Figures 9 and 10, are markedly similar.

DISCUSSION

Overall, the Post-it™ notes provided a useful, convenient method for collecting dust particles for qualitative analysis with the scanning electron microscope. It appears to be most useful for particles larger than 0.1mm, although 1 micrometer (μm) particles containing heavy elements are readily detected with the backscatter electron imaging capability.

Because the adhesive is somewhat electron beam sensitive (softening under the heat of the beam), smaller particles may sink below the surface during analysis. The background of zinc/chlorine-containing particles will interfere with some elemental analyses, but do not interfere significantly with analyses for asbestos or heavy elements of environmental interest such as lead, cadmium or chromium.

The Post-it™ is most useful in areas of light dust where it is difficult to gather sufficient quantities to be placed in a container.

Overloading of Post-it™ notes may be a problem as there is a tendency on the part of the collector to sample very dusty areas. Ideal sampling with

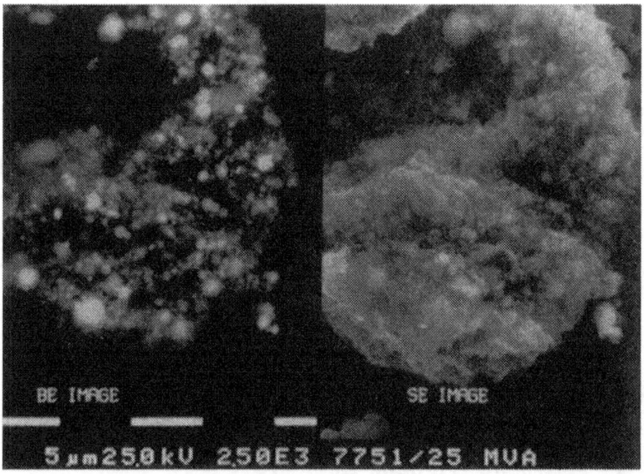

Figure 10. Backscatter electron image (L) and secondary electron image (R) of fly ash particles from in the house. Bar = 5µm.

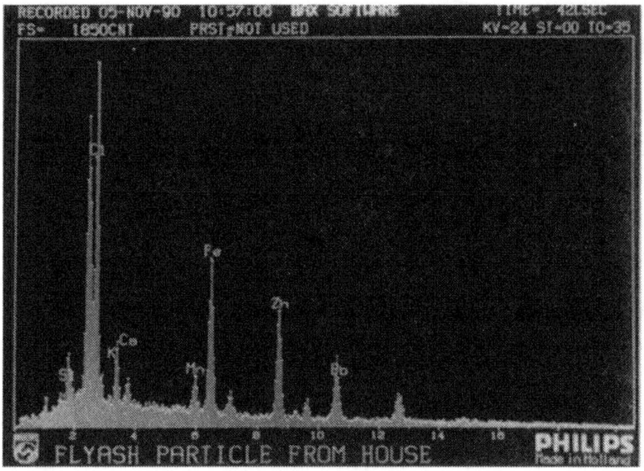

Figure 11. X-ray spectrum of a fly ash particle in Figure 10.

the Post-it™ note would have a moderate amount of particulates adhering to the glue strip, but not so many that the Post-it™ could not be folded over onto itself forming a seal.

CONCLUSION

Post-it™ notes are a very convenient media for sampling dust for qualitative SEM analysis. They are readily available and the collection technique is easily performed and can be described over the telephone to the collector. Analysis of particles in the dusts collected on Post-it™ notes may be useful in determining the source of the dust.

REFERENCES

1. Hopen, T. Personal Communication. 1989.
2. Fisk, J.T. "Tricks of the Trade, 3M's 'Post-it™' as an Aid in Particle Sorting." *Microscope.* 38:197–198. 1990.

NOTES

™ 3M Company, St. Paul, MN 55144

AUTHOR AFFILIATIONS

* MVA, Inc., Norcross, GA
** Global Site Analysis, Inc., Atlanta, GA
*** Azimuth, Inc., Norcross, GA

Source: NAC Journal, Spring 1991. Used with permission of The Environmental Information Association, Atlanta, GA.

APPENDIX 10

Exposure to Airborne Asbestos Associated with Simulated Cable Installation Above a Suspended Ceiling*

D.L. Keyes[a], J. Chesson[b]†, W.M. Ewing[c], J.C. Faas[a], R.L. Hatfield[d], S.M. Hays[e], W.E. Longo[f], J.R. Millette[g]‡

Installing cable above a suspended ceiling in the presence of asbestos-containing fireproofing is an example of an activity that may disturb in-place asbestos and associated dust and debris. Two simulations of cable installation were conducted in a room of an unoccupied school to test the extent of such disturbance and resulting elevations in airborne asbestos. Average airborne asbestos concentrations in the room increased over 500-fold during the simulations, with several samples exceeding 50 structures per cubic centimeter (s/cm³), as measured by transmission electron microscopy (TEM) with an indirect preparation technique. Elevated concentrations persisted during a subsequent cleaning of horizontal surfaces in the room and for several hours thereafter. Personal samples collected on the cable installers yielded TEM measurements averaging approximately 68 s/cm³ for the two simulations.

Several studies have indicated that maintenance, repair, and construction activities in buildings with asbestos-containing material (ACM) can disturb the ACM and raise levels of airborne asbestos.[1–3] Installing electrical or computer cable above a suspended ceiling in the presence of asbestos fireproofing is a typical example of such activities. In-place fireproofing may be accidentally contacted and asbestos-containing dust and debris accumulated on ceiling tile may be disturbed. Resulting release or re-entrainment of asbestos fibers will likely lead to elevated levels of airborne asbestos. Although quantitative evidence of these effects is limited, a few studies have shown airborne fiber concentrations measured by phase contrast microscopy (PCM) to be over 50 fibers per cubic centimeters (f/cm^3).[2] Most of these studies are not well characterized with respect to the type and duration of activities and the duration of elevated air levels. Moreover, transmission electron microscopy (TEM), rather than PCM, is now recognized as the method of choice to analyze air samples for asbestos.[4–5] TEM can distinguish between asbestos and non-asbestos fibers and can detect thin fibers that cannot be resolved by PCM. For these reasons, a controlled study of cable installation in the vicinity of asbestos fireproofing was undertaken.

The study was designed specifically to examine the effects of normal cable installation procedures without regard to the presence of asbestos. Accordingly, special precautions to minimize asbestos emissions were not taken. The study thus examines situations where workers do not know about the presence of asbestos or where they do not either know or use work practices specially designed for work around asbestos. The authors' experiences working with owners of hundreds of buildings with ACM has indicated that these situations are not uncommon. The study results provide a rationale for using special work practices and can serve to establish a baseline against which the effectiveness of these practices can be measured.

TEST SITE

The study was conducted in a 20-year-old unoccupied school awaiting removal of the asbestos-containing fireproofing (Aspen Middle School in Aspen, CO). A room used for science education was selected as the test site (Figure 1). It contained a large central space, a separate area in one corner created by partitions extended to the suspended ceiling, and two office areas and two storage areas separated by masonry walls extending to the deck above. Total floor area was approximately 2000 ft^2. Floor coverings were carpet and vinyl tile. All movable furnishings had been removed before the start of the study.

Most of the room contained a suspended ceiling of 61 × 122 cm (2 × 4 ft) lay-in, asbestos-free tile; one office contained 30.5 × 30.5 cm (1 × 1 ft) hidden spline tile. Friable fireproofing with 15–20% chrysotile asbestos, 45–55% vermiculite, and 20–40% binder was sprayed on steel beams with overspray on the deck and utilities above the suspended ceiling. The fireproofing generally was in good condition. Some dust and debris similar in appearance to the fireproofing was visible on top of the ceiling tile. The space between the ceiling tile and the deck above was approximately 0.9 m (3 ft) in height and served as a return air plenum during normal building

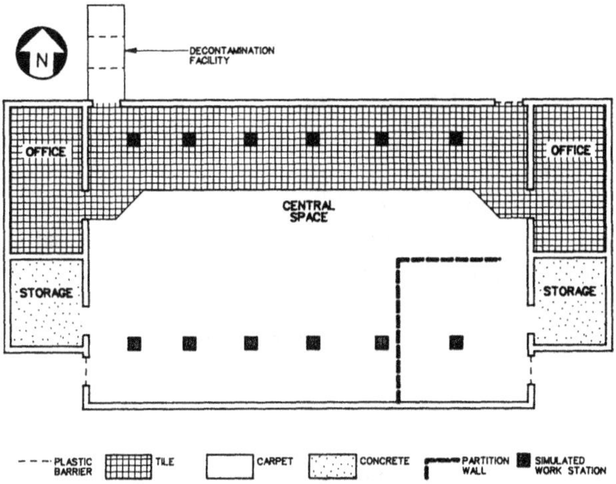

Figure 1. Diagram of the test site.

operations. This space had been entered a few times per year for utility system maintenance during the life of the building.

The room was prepared for the study by covering the doors and wall penetrations with 6-mil plastic sheets and duct tape. The HVAC system was shut down and was to be demolished and replaced as part of the subsequent asbestos abatement project. A three-chamber decontamination facility for personnel and equipment was constructed at one entry to the room, as shown in Figure 1.

STUDY DESIGN

Two identical simulations were performed. Each consisted of four phases: cable installation above the suspended ceiling (approximately 2.5 hr); an inactive period (approximately 1.5 hr); cleaning of horizontal surfaces in the room with a commercial vacuum cleaner, broom, and dry cloths (approximately 1 hr); and a final inactive period (about 3 hr). At the end of each 8-hr simulation, wet cloths and a high-efficiency particulate air (HEPA) vacuum cleaner were used to clean the floor and other horizontal surfaces and two HEPA air filtration units were used to clean the air.

To measure the effect of cable installation and subsequent cleaning of the test site, air samples were collected before each simulation began, during the installation activity, and at various times during the 6.5-hr postinstallation period. The experiment was designed to detect 10-fold or greater differences in average airborne asbestos levels among these periods with a probability of at least 95%. Assuming that the coefficient of variation for the TEM asbestos measurements would be approximately 100%, the number of samples was determined to be eight per period. Because the differences were actually much larger than 10-fold, it was not necessary to analyze all samples.

The quality assurance program included side-by-side samples and field and laboratory blanks. In addition, duplicate analyses of selected filters were performed by independent laboratories.

EXPERIMENTAL MATERIALS AND METHODS

Simulation Procedures

Cable was installed by using procedures typical of those used in the trade. From the northwest corner of the room, 12 separate cables were run to 12 simulated computer work stations, as illustrated in Figure 1. The simulation began by removing ceiling tile above each "work station" and as needed along the route each cable would trace above the suspended ceiling. Altogether, about 15 out of roughly 200 individual tiles were moved. The removed tiles were then stacked on the floor. Working from ladders, installers attached a sting weighted with a 2.54-cm (1-in.) nut to one end of each cable. The nut was then thrown in the direction of the appropriate work station. A second installer located the nut and pulled the sting and cable to that location. This "throw and retrieve" operation was used to bring a cable to each work station. Typically, one or two throws were required for each retrieval. In the process, in-place fireproofing was struck about five times during the process of installing 12 cables. Figure 2 illustrates the installation process. All cables were then pulled back to the starting point, recoiled, and set aside, and all ceiling tiles were replaced. The entire operation of installing, removing, and recoiling cable is essentially equivalent to a 2.5-hr period of installation.

The cleaning operation consisted of vacuuming all carpeted areas; wiping all lab benches, window sills, and other horizonal surfaces with a dry cloth; and sweeping the tiles portion of the floor. No attempt was made to suppress re-entrainment of dust.

Equipment and Data Collection Methods

Air-sampling equipment used during the study included high-flow sampling pumps (Dawson Associates, Tucker, GA) and personal sampling

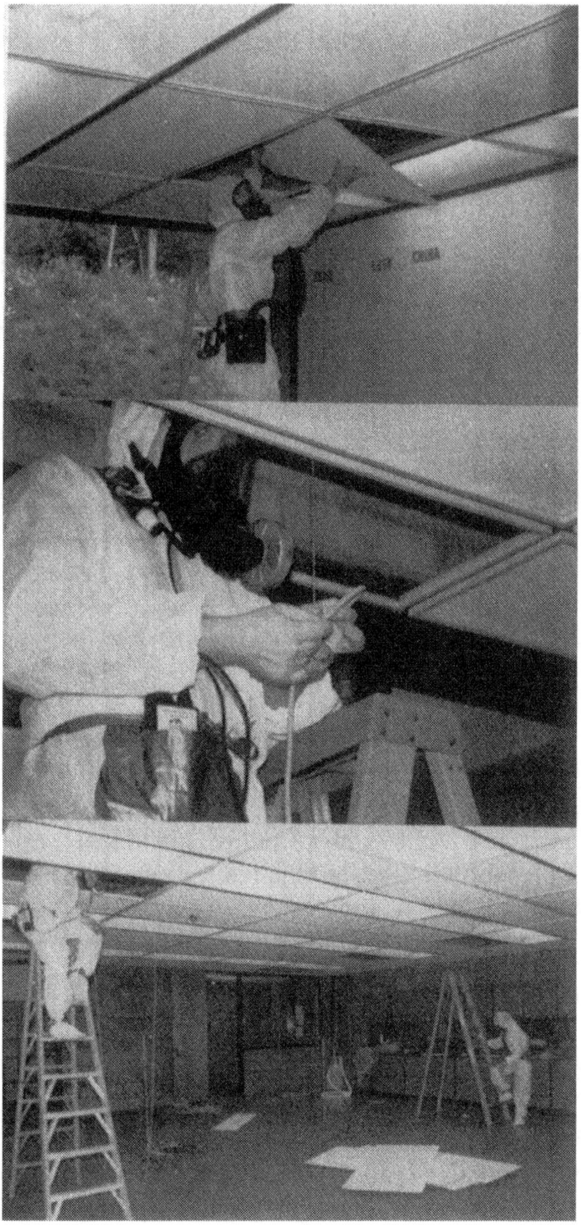

Figure 2. Cable installation process.

pumps (Mine Safety Appliance Company, Pittsburgh, PA). Flow measurements were performed on-site with a Mini-Buck™ Model M-30 automated soap bubble calibrator (A.P. Buck, Orlando, FL). The sampling media consisted of 0.45-µm pore size mixed cellulose ester filters housed in 25-mm sampling cassettes (Nuclepore Model SN32275, Cambridge, MA). Other equipment used during the study included a HEPA vacuum cleaner (Hako Minuteman™, Addison, IL). HEPA air filtration units with nominal flow rates of 1950 ft^3/min (Model Mach 2, Critical Systems, Houston, TX), a Dial-a-matic™ upright vacuum cleaner (Hoover Vacuum Cleaners, North Canton, OH), and various personal respirators. Number 16/3 electrical cord was used as the computer cable.

The flow rate of each air-sampling pump was measured before and after each experiment. High-flow pumps were placed at various locations both inside and outside the room, as shown in Figure 3. Personal pumps were placed on selected workers. Airflow rates were approximately 5.4–8.1 L/min for the high-flow pumps and 1.9 L/min for the personal pumps. Overloading of filters during the simulations required changes of cassettes on each pump. All cassettes were taped closed and transported by one of the authors to the laboratories for analysis.

Asbestos Analysis

Subsets of area and personal samples were selected to reflect a wide spatial distribution within the room. Each of the selected samples was analyzed by TEM. Because the filters were heavily loaded, they were prepared for analysis by TEM with an indirect rather than a direct technique. Direct sample preparation, such as required for asbestos abatement projects in schools (Asbestos Hazard Emergency Response Act—40 CFR 763), is appropriate for measuring low concentrations of airborne asbestos in relatively clean environments. A quantitative measure of the limitations of direct TEM preparation is provided by the National Institute for Occupational Safety and Health (NIOSH) for Method 7402—the upper boundary of airborne asbestos concentrations that can be reliably measured is 0.5 f/cm^3.[6] Where asbestos concentrations are high or the air is heavily contaminated with non-asbestos particles, obscuration of asbestos fibers by other fibers or by non-asbestos particles reduces the reliability of fiber counting.[7–8] Indirect sample preparation includes filter ashing to remove organic materials, followed by water suspension, sonication to mix the ashed material, and refiltering. Suspension, sonication, and refiltering allows for dilution of the sample as needed and more even distribution of fibers for TEM viewing, both of which improve the reliability of fiber counting.[7–8]

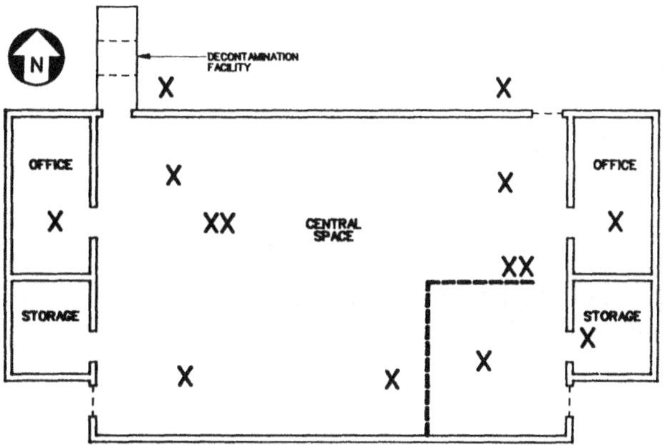

Figure 3. Location of stationary air-sampling pumps.

The indirect method of sample preparation and microscope counting rules as described in the Yamate method, Level II[9] were employed with one exception—sonication time was 10 min rather than 3 min. Either one-fourth or one-eight of each filter (depending on the degree of dust loading) was ashed in a plasma asher. The ashed material was then suspended in distilled water, sonicated, refiltered, and mounted for TEM examination by using the Jaffe-Wick method. All asbestos structures that could be resolved were reported.

Personal samples were also analyzed by PCM. NIOSH Method 7400 was used with the "A" counting rules.[10]

Statistical Analysis

Where there were two or more consecutive samples at a particular location during a given sampling period, a time-weighted average was calculated so that all measurements of airborne asbestos concentrations are based on equivalent time periods. Statistical analyses were applied to the logarithm (Base 10) of the measured concentration. The log transformation tends to equalize variances and permit the use of standard statistical tests that would otherwise be inappropriate. Previous studies of indoor air pollution have demonstrated that air pollution data tend to be lognormally distributed.[11]

The effect of "sampling period" as a variable was tested with a standard one-way analysis of variance and pairwise comparisons between individual sampling periods were performed by using the Scheffé test.[12] Results are expressed as p-values. The smaller the p-value, the stronger the evidence for a difference between the sampling periods being compared. P-values less than 0.05 are usually regarded as statistically significant.

Ratios or airborne asbestos concentrations and their 95% confidence intervals were estimated by calculating the ratio of the corresponding sample geometric means. Geometric as opposed to arithmetic means are the appropriate quantities to use to estimate the ratio when the distributions are lognormal and the coefficients of variation are equal.

RESULTS

Table 1 shows the results for the area samples for the first simulation. The average concentration of airborne asbestos (as measured by TEM) during the cable installation operation is estimated to be 590 times higher than the concentration before the experiment began (geometric means of 26.9 and 0.046 s/cm^3). The 95% confidence interval (CI) for this ratio is 230 to 1500. Likewise, the average concentration during the postinstallation period is 140 times higher than before the experiment began (CI = 50 to 330). Both these differences have a p-value (p) of less than 0.0001. The concentration during cable installation is over four times higher than during the postinstallation period (CI = 1.8 to 11, p = 0.01).

Table 2 reveals similar patterns for the second simulation. Average airborne asbestos concentrations during cable installation are 520 times higher (67.4 versus 0.129 s/cm^3) than those before the experiment began (CI = 140 to 1950, $p < 0.0001$) and those during the postinstallation period are 95 times higher than those before (CI = 25 to 360, p = 0.0001). The ratio of average concentration during cable installation to average concentration during the postinstallation period is over 5 (CI = 1.4 to 22, p = 0.12), but the difference is not statistically significant at the 0.05 level.

Tables 1 and 2 also show the TEM results for the personal samples during cable installation for each of the two simulations. The relationship between area and personal samples differs between simulations. In the first simulation, the average concentration measured by area samples during installation is higher than the concentration measured by personal samples (p = 0.04). In the second simulation, the average concentration measured

Table 1. Descriptive statistics of TEM results for the first simulation.

PHASE	ARITHMETIC MEAN (s/cm^3)	ARITHMETIC STD. DEV. (s/cm^3)	GEOMETRIC MEAN	NUMBER OF OBSERVATIONS
Before installation	0.052	0.030	0.046	5
During installation (area samples)	28.9	12.6	26.9	5
During installation (personal samples)	10.5	11.6	7.1	3
After installation	8.4	7.0	6.2	6

Table 2. Descriptive statistics of TEM results for the second simulation.

PHASE	ARITHMETIC MEAN (s/cm^3)	ARITHMETIC STD. DEV. (s/cm^3)	GEOMETRIC MEAN	NUMBER OF OBSERVATIONS
Before installation	0.158	0.094	0.129	5
During installation (area samples)	100.2	91.9	67.4	4
During installation (personal samples)	124.8	85.6	102.7	3
After installation	17.0	13.5	12.3	4

by area samples during installation is lower, but the difference is not statistically significant at the 0.05 level ($p = 0.6$). Nevertheless, both the area and personal samples are significantly elevated compared with the levels prior to cable installation.

Figure 4 summarizes the results for the two simulations. Note that the vertical axis is on a logarithmic scale. The average levels of airborne asbestos (geometric means, or means on the log scale) are represented by the height of each bar for each of the three phases of the experiments. The variation about the mean (standard deviation on the log scale) is indicated by the vertical lines. Also shown are the means and standard deviations of the personal samples during the installation and postinstallation periods.

A few personal samples collected during cable installation were also analyzed by PCM for comparison purposes. The arithmetic means are 0.13 f/cm^3 (SD = 0.03) for the first simulation and 0.34 f/cm^3 (SD = 0.11) for the second simulation.

The quality assurance results indicated that field and laboratory contamination was minimal with one exception. Of the 13 field and laboratory blanks analyzed, 1 field blank showed a higher than expected structure loading (441 s/mm^2). The laboratory blanks averaged 35 s/mm^2 and the field blanks averaged 151 s/mm^2.

Initial comparison of side-by-side samples analyzed by two independent laboratories showed substantial differences in results. Investigation of these results suggests that differences in sample preparation and analysis—time of sonication, size of the filter used for refiltration of the ashed sample, and failure of one laboratory to count small asbestos structures (less than 0.5 µm in length)—accounts for many of these differences, in addition to any inherent difference in the air sampled between side-by-side samples. To further test this conclusion, the quality assurance laboratory reanalyzed a different portion of the same filter (to avoid differences between side-by-side samples) for six stratified random samples, holding constant the sonication time, filter size, and structure counting variables. Although the second laboratory produced results that were approximately half as large as those of the first laboratory, there was a high and statistically significant correlation (0.88, $p = 0.02$) between the two sets of results.

DISCUSSION

The results show that airborne asbestos concentrations as measured by TEM are significantly elevated during cable installation. Concentrations generated during the study were high both relative to concentrations before

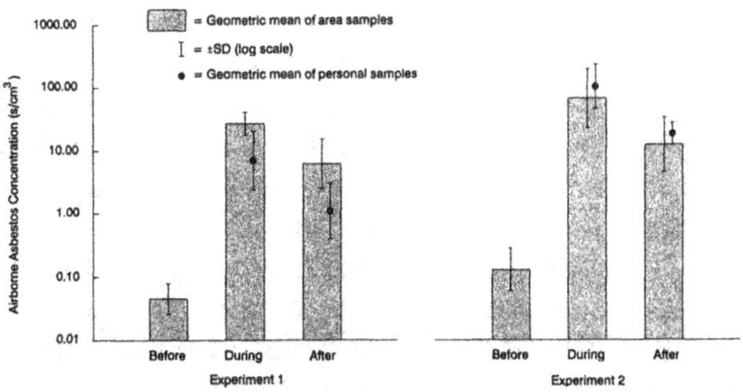

Figure 4. Summary of TEM results.

the experiment and in absolute terms. Concentrations peaked during installation. Whether cleaning activities resuspended asbestos structures and thereby contributed to the elevated levels after cable installation cannot be determined directly from the data. However, elevated concentrations persisted for several hours after the installation was finished.

For the personal samples, asbestos concentrations measured by TEM are substantially higher than might be expected from the fiber levels recorded by PCM. This suggests that most of the measured asbestos structures are thinner than the resolution limit of PCM or shorter than those fibers counted according to NIOSH Method 7400. An examination of fiber size distributions reported for the TEM analyses verifies this conclusion—over 95% of measured structures are thinner than 0.25 µm and shorter than 5 µm.

Health scientists continue to debate the relationship between fiber size and potency; the arguments have recently been summarized by Chesson et al.[13] The current Occupational Safety and Health Administration (OSHA) exposure standards are based on measurements that use NIOSH Method 7400 (or equivalent) and thus on fibers longer than 5 µm and with diameters greater than about 0.25 µm. However, these fiber dimensions do not reflect an official OSHA view of the health question; rather, they derive from historical considerations of fiber counting reliability.[14] OSHA recognizes that PCM is neither specific for asbestos nor sensitive to small (both short and thin) fibers; in this light, PCM measurements are best viewed as an indicator of exposure to airborne asbestos. How adequate an indicator PCM measurements are depends in part on the health significance of the small, unmeasured fibers. Research continues on the question.

The analytical method used in the present study measures asbestos structures of all sizes. The term "structures" is used to identify free asbestos fibers, overlapping fibers identified as "bundles" and "clusters," and fibers combined with matrix materials called "matrices." Moreover, sonication of ashed samples may disassociate bundles and clusters and disassemble matrices, thereby increasing the total number of structures counted.[15] (Some have suggested that sonication may actually break individual asbestos fibers but this hypothesis does not accord with available evidence.[15])

The significance of the elevated concentrations of airborne asbestos recorded in the present study will depend on the physiological importance of the asbestos structures identified and counted. One could argue that the disassociation and disassembly of complex structures during sonication mimics natural physiological processes in the human lung.[16] Thus, indirect preparation of TEM samples not only improves fiber counting reliability, it may provide a better measure of asbestos fibers presented to the lung. However, research is needed to substantiate this possibility.

Despite the remaining uncertainty about the significance of small asbestos fibers and the disassembly of complex structures during indirect sample preparation, cable installation workers are well advised to wear respirators and to use work practices designed to reduce fiber release and resuspension. The recorded concentrations of airborne asbestos are high even if only 5% of the structures are ultimately shown to have health consequences. The persistence of high asbestos levels during the simulation suggests that work areas should be isolated from the rest of the building and the air cleaned with air filtration units before the areas are reoccupied, unless use of wet methods and good work practices can greatly reduce concentrations of airborne asbestos.

The extensive quality assurance aspects of this study identified two potential problems. The first involves possible contamination of samples. The one blank sample with an elevated reading (441 s/mm^2) suggests that contamination may have contributed to the observed levels of airborne asbestos. However, only one of three field blanks showed elevated loadings and none of the background samples showed unexpectedly high levels. Moreover, even if contamination was widespread, a structure loading of 441 s/mm^2 is low compared to the loading on all but the background samples (the average loading was approximately 38,000 s/mm^2 for samples collected during the simulation.) Thus, contamination is not believed to be a problem.

The second potential problem concerns interlaboratory differences in analytical results. Although the investigation identified the major causes of the discrepancies, the quality assurance laboratory's results were still approximately one-half as large as the primary laboratory's results. However, the high correlation between the two sets of results means that the study's conclusions would be unchanged regardless of which laboratory analyzed the samples. Systematic differences between laboratories is probably not unusual for TEM analysis of asbestos in air; the authors' observations suggest that studies involving airborne asbestos measurements should utilize data from more than one laboratory. Other investigators are encouraged to conduct similar quality assurance examinations and to probe the causes of any observed interlaboratory differences.

ACKNOWLEDGMENT

The authors wish to thank the Aspen School District and its employees, particularly Mr. Griff Smith, for their cooperation. Brand Site Services and Kaselaan & D'Angelo Associates also assisted in providing materials, equipment, and logistical support.

REFERENCES

1. **Sawyer, R.N.:** Asbestos Exposure in a Yale Building. *Environ. Res.* 13:146–169 (1977).
2. **Paik, N.W., R.J. Waicoti and P.B. Pairicia:** Worker Exposure to Asbestos During Removal of Sprayed Material and Renovation Activity in Buildings Containing Sprayed Material. *Am. Ind. Hyg. Assoc. J.* 44(6):428–432 (1983).
3. **Pielart, J.H. and R.G. Mathey:** *Guidelines for Assessment and Abatement of Asbestos-Containing Materials in Buildings.* Washington, DC: General Sciences Administration, 1983.
4. **U.S. Environmental Protection Agency:** *Measuring Airborne Asbestos Following an Abatement Action* (EPA 600/4-85-049). Washington, DC: U.S. Environmental Protection Agency, 1985. pp. 3-1–3-7.
5. **Stewart, I.M.:** Asbestos—Analytical Techniques. *Appl. Ind. Hyg.* 3(6):F-24–F-27 (1988).
6. **National Institute for Occupational Safety and Health:** *NIOSH 7402 Method Protocol.* Washington, DC, National Institute for Occupational Safety and Health, 1987.
7. **Sebastian, P.:** "Direct or Indirect Transmission Microscope Methods for Measuring Asbestos Air Pollution in the Environment." Paper presented at the American Industrial Hygiene Conference, Orlando, FL, May 1990.
8. **Chatfield, E.J.:** Asbestos Measurements in Workplaces and Ambient Atmospheres. In *Electron Microscopy in Forensic Occupational and Environmental Health Sciences,* edited by S. Basu and J.R. Millette. New York: Plenum Publishing Corporation, 1986.
9. **Yamate, G., S.C. Agarwal and R.D. Gibbons:** *Methodology for the Measurement of Airborne Asbestos by Electron Microscopy.* Research Triangle Park, NC: U.S. Environmental Protection Agency, 1984 (Draft). Copies can be obtained from Mr. Michael Beard at U.S. EPA (919-541-2623).

10. **National Institute for Occupational Safety and Health:** *NIOSH 7400 Method Protocol.* Washington, DC: National Institute for Occupational Safety and Health, 1984.
11. **Gilbert, R.O.:** *Statistical Methods for Environmental Pollution Monitoring.* New York: Van Nostrand Reinhold, 1987. p. 152.
12. **Miller, R.G.:** *Simultaneous Statistical Inference.* New York: Springer-Verlag, 1981.
13. **Chesson, J., J.D. Rench, B.D. Schultz, and K.L. Milne:** Interpretation of Airborne Asbestos Measurements. *Risk Anal.* 10(3):437–447 (1990).
14. "USDOL/OSHA Occupational Exposure to Asbestos, Tremolite, Anthophyllite, and Actinolite: Final Rules." *Federal Register* 51:119 (20 June 1986). P. 22,680.
15. **U.S. Environmental Protection Agency:** *Comparison of Airborne Asbestos Levels Determined by Transmission Electron Microscopy (TEM) Using Direct and Indirect Transfer Techniques* (EPA 560/5-89-004). Washington, DC: U.S. Environmental Protection Agency. 1990.
16. **Timbrell, V.:** Review of the Significance of Fiber Size in Fiber-Related Lung Disease: A Centrifuge for Preparing Accurate Microscope-Evaluation Specimens from Slurries Used in Inoculation Studies. *Ann. Occup. Hyg.* 33(4):483–505 (1989).

NOTES

* This work was supported in part by the Plaintiff's Executive Committee of the National Schools' Class Action for Cost Recovery; the authors exercised full control over design and execution of the study and the interpretation and reporting of study results. A paper summarizing the work was presented at the American Industrial Hygiene Conference in Orlando, FL, May 1990.

† Dr. Chesson is currently with the Australian Resource Assessment Commission, Queen Victoria Terrace, Canberra, ACT 2600, Australia.

‡ Dr. Millette is currently with Millette, Vander Wood and Associates, Inc., 5500 Oakbrook Parkway, Suite 200, Norcross, GA 30093.

AUTHOR AFFILIATIONS

[a]Environmental Sciences, Inc., 105 E. Speedway Blvd., Tucson, AZ 85705; [b]Chesson Consulting, Inc., 1717 Massachusetts Ave., NW, Suite 601, Washington, DC 20036; [c]Diagnostic Engineering, Inc., 350 Franklin Rd., Suite 300, Marietta, GA 30067; [d]Law Associates, 14 Town Park Dr., Kennesaw, GA 30144; [e]Gobbell Hays Partners, Inc., 217 Fifth Avenue, N., Nashville, TN 37219; [f]Materials Analytical Service, 3597 Parkway Ln., Suite 250, Norcross,

GA 30092; [g]McCrone Environmental Services, Inc., 1412 Oakbrook Dr., Suite 100, Norcross, GA 30013

Source: American Industrial Hygiene Association Journal. Volume 52. November, 1991. Used with permission, American Industrial Hygiene Association, Fairfax, VA.

Index

For Product Safety Concerns and Information please contact our EU representative GPSR@taylorandfrancis.com
Taylor & Francis Verlag GmbH, Kaufingerstraße 24, 80331 München, Germany

www.ingramcontent.com/pod-product-compliance
Lightning Source LLC
Chambersburg PA
CBHW070824150926
51501CB00043B/974
9780367579784